To Master the Boundless Sea

**Flows, Migrations, and Exchanges**

Mart A. Stewart and Harriet Ritvo, *editors*

The Flows, Migrations, and Exchanges series publishes new works of environmental history that explore the cross-border movements of organisms and materials that have shaped the modern world, as well as the varied human attempts to understand, regulate, and manage these movements.

# To Master the Boundless Sea

The U.S. Navy, the Marine Environment, and the Cartography of Empire

· · · · · · · · · · · · · · · · · · · · · · · · · · · · · · · · · · · · · · · · · · ·

JASON W. SMITH

University of North Carolina Press  Chapel Hill

*This book was published with the assistance of the Anniversary Fund of the University of North Carolina Press.*

The University of North Carolina Press has been a member of the Green Press Initiative since 2003.

Library of Congress Cataloging-in-Publication Data
Names: Smith, Jason W. (Jason Wirth), 1982– author.
Title: To master the boundless sea : the U.S. Navy, the marine environment, and the cartography of empire / Jason W. Smith.
Description: Chapel Hill : University of North Carolina Press, [2018] | Includes bibliographical references and index.
Identifiers: LCCN 2017045197| ISBN 9781469640440 (cloth : alk. paper) | ISBN 9781469659220 (pbk. : alk. paper) | ISBN 9781469640457 (ebook)
Subjects: LCSH: United States. Navy—History. | United States—History, Naval. | Nautical charts—United States—History. | Cartography. | Imperialism and science.
Classification: LCC E182 .S574 2018 | DDC 359.00973—dc23
LC record available at https://lccn.loc.gov/2017045197

Cover illustration: *Hove-to for a Pilot* (1893) by Henry Moore (Image BHC4175, National Maritime Museum Picture Library, © National Maritime Museum, Greenwich, London).

Portions of this book were previously published in a different form and are used here with permission. Material from chapter 2 appeared in "The Bound[less] Sea: Wilderness and the United States Exploring Expedition in the Fiji Islands," *Environmental History* 18 (October 2013): 710–37. Material from chapter 3 appeared in "Matthew Fontaine Maury: Pathfinder," *International Journal of Maritime History* 28 (May 2016): 411–20, published by SAGE Publications Ltd., all rights reserved, © Jason W. Smith. Material from chapter 5 appeared in " 'Twixt the Devil and the Deep Blue Sea: Hydrography, Sea Power, and the Marine Environment, 1898–1901," *Journal of Military History* 78 (April 2014): 575–604.

*To my parents and grandparents, my students,*
*and, above all, to Megan and Nel, with love and gratitude*

Queequeg was a native of Kokovoko, an island far away to the West and South. It is not down in any map, true places never are.

—Herman Melville, *Moby-Dick*

# Contents

# Illustrations

# Introduction

## The Boundless Sea

· · · · · · · · · · · · · · · · · · · · · · · · · · · · · · · · · · · · · · · · · · · · ·

Nothing like a lead-line well swung . . . to sharpen an agile mind.
—Robert Seager II, *Alfred Thayer Mahan: The Man and His Letters*, 1977

In January 2013, the minesweeper USS *Guardian* struck a coral reef in the Philippines. Precariously listing, pounded by surf against coral, the ship was a total loss. The incident was an embarrassment for the navy and the United States, another moment in a long, complicated history with the Philippines. Filipinos and others openly wondered what the vessel was doing there in the first place, recalling the United States' long imperial shadow in the islands going back to the Spanish-American War of 1898. The navy blamed inaccurate digital charts in attempting to explain why a modern warship, equipped with charts, sonar, and other navigational technologies, could find itself hard aground—an unlikely flash point for questions of empire, navigation, cartography, science, and nature.[1]

As historian Alistair Sponsel has observed, perhaps the most interesting aspect of the *Guardian*'s fate was not the grounding itself, but the collateral damage to the natural world. The vessel struck Tubbataha Reef, eighty miles east of the island of Palawan, a UNESCO World Heritage Site, home to one of the most pristine coral reef environments in the world, and protected under Philippine law. Through the winter and spring of 2013, the $277 million vessel was scrapped, not by pulling it off the coral, which would have done additional damage to the ecosystem, but by tedious, piecemeal disassembly on the reef as salvors' torches and saws cut the ship apart, in effect, finishing what the natural environment had begun. Coral reefs, Sponsel argues, have been transformed from dangerous to endangered. For thousands of years, mariners hoped to save their ships from them. Now, it seems, we desire to save reefs from ships.[2]

The *Guardian*'s fate underscores a number of important questions taken up in the following pages. Ships run aground, and when they do, vessels are damaged or lost, and nations and their people are left to wonder about imperial relationships and the fragility and power of nature. Hydrographic charts and other navigational tools, no matter their sophistication, are often flawed representations of a complex, dynamic, natural world whose presence—today and historically—is a pervasive element of naval operations. What role does the marine environment play in navigation, strategy, and empire more broadly? Directly or indirectly, the *Guardian*'s grounding raised these issues anew. Of course, such questions are as old as navies themselves. From its origins in the 1790s, the U.S. Navy has struggled to understand, to use, and sometimes to control and change the marine environment to further the nation's commercial and strategic interests. This book tells part of that story.

It does so by focusing on American naval scientists, whose work surveying, charting, and thus representing and imagining the marine environment through the nineteenth century played a critical and often overlooked role in the expansion of American empire. The navy's hydrographers and the charts they produced provide a lens with which to more deeply examine the historical encounter between navies and the marine environments that so fundamentally shape their activities. It also explains the growth of American power and the expanse of empire in ways that transcend the usual manifestations of military power on land and sea—battles won and treaties signed—for the subtler, but no less potent, tools of science and cartography.

What follows is a watery history of American empire. It is set in the liminal spaces where surf meets sand beach, coral reef, or ice-bound shore, from the deep sea to a littoral often labyrinthine in navigational hazards, from the mysteries of the deep ocean to the bays, harbors, and coastlines that together formed the natural spaces of American oceanic empire in the nineteenth century. From the treaties, battles, and other discourses that collectively define the historiography of American empire in the nineteenth century, this work descends to the water itself, examining the nuts and bolts of empire building in a marine environment that was central but often hostile or indifferent to such intentions. Outside the continental United States, in waters where the nation sought to extend first its commercial might, and later, its more muscular visions of "sea power"—to use a phrase coined by the naval officer and theorist Alfred Thayer Mahan in 1890—American naval officers used hydrography, broadly defined as the study of ocean depths, winds, currents, tides, and meteorology, to claim, recast, and control the marine environment. In doing so, they professed a scientific faith in the empiricism and precision of their methods that was just as imperial in its vision and execution as other more obvious forms of American power at sea. At sea, the empirical was inextricably imperial.[3]

In the nineteenth century, American ocean empire was not primarily territorial; it did not, generally speaking, play out within the political and legal framework that we often ascribe to the emergence of formal empire at least until the late nineteenth century, when the United States began to possess its own. Even after the 1898 Spanish-American War, the Americans' cartographic and strategic claims to the sea were amorphous, elusory, and contested. Still, they remained potent ways for articulating a vision of American empire that had tremendous implications for the coming century. Rather than territorial claims, the chart's power rested in the ways it could represent, frame or recast, and more broadly visualize the ocean for commercial navigators, naval officers, and naval scientists intent on constructing a cartographic ocean that reflected the nation's growing commercial and strategic interests. From its birth, the United States has generally been committed to the principles of freedom of the seas articulated by the Dutch jurist Hugo Grotius in his *Mare Liberum* published in 1608. Grotius argued, "The sea is common to all . . . it cannot become a possession of any one." Nevertheless, the United States staked subtler claims to the ocean that extended well beyond the three-nautical-mile limit of American territorial waters.[4]

Charts were malleable and versatile, if also problematic, instruments of power. As navigational aids and in a larger thematic cartography that transcended the navigational chart's soundings to represent winds, currents, ship tracks, and, later, strategic representations of nature, charts imposed meaning on waters newly important to American trade and defense. They set down the routes through which the veins of maritime commerce flowed and, more importantly, they mediated mariners' and naval officers' encounters with the sea, assisting them in visualizing and orienting their place on what many otherwise considered a "boundless" ocean. Claims of precedence and the appropriation of ocean waters followed the surveyors' and cartographers' wake—sometimes in political and legal form—but more often in cultural and imagined terms. The sea, by nature, often eschewed terrestrial structures of ownership or control such as fenced land, political borders, or legal constructs. How do nations stake claims to transnational spaces like the ocean? How do they control the great common, to borrow another line from the naval theorist Mahan? Hydrography, in other words, laid the foundation on which the United States attempted to build its oceanic empire outside the territorial waters of the United States. It was far from the only factor, of course, but it was significant, and it has been understudied.

Often, though, hydrographers and surveyors struggled with a marine environment that was vast, deep, and in constant flux. The sea could render precise scientific measurements difficult and sometimes impossible. Where sea met land, the natural environment was awash with movement. Sandbars shifted. Coastlines were made and remade. The deep sea presented challenges to human knowledge that technology could not always overcome. It continually flouted human pretensions to represent it. The difference between the two-dimensioned chart, braced as it was by the graticule of latitude and longitude, and quantified in depth soundings on the one hand, and on the other, the physical environment in all its ebbs and flows, cragged coral heads, and foaming breakers, was vast. In the disparities between the chart and the physical environment it purported to represent, we can better understand both the expanse and the limits of American empire at sea.

*To Master the Boundless Sea* places the sea at the center of the historical narrative, borrowing from the methods and historiographies of military, naval, and maritime history, the history of science and cartography, environmental history, cultural history, and the history of technology to construct a new interpretation concerning the role of science and the marine environment in the expansion of empire and the growth of American military power. I begin with the emergence of American naval science in the 1830s, trace the evolution of hydrographic science in antebellum exploring expeditions and the work of the Naval Observatory and Hydrographic Office in Washington, D.C., before moving into a post–Civil War ocean changed—first slowly and subtly, but later with greater urgency—by new strategic understandings that transformed both the physical environment and the way the navy imagined it. I conclude by examining the environmental and hydrographic challenges of naval war with Spain in 1898 and the subsequent rise of charts and a broader thematic cartography in influencing the outcome of important strategic debates about American power from the Caribbean to the western Pacific in the first years of the twentieth century. The structure is episodic. I neither document every survey made, nor each chart constructed. This is not an institutional history of naval science, which encompassed both astronomy and hydrography, and, it might be added, naval architecture, gun design, and engineering as well. Rather, the book focuses on several important, meaningful moments throughout the nineteenth century that place the transformation of maritime and naval power among the reefs, rocks, shoals, and depths on which the American ocean empire rested and whose features the sea chart sought to capture.

In seeking to recover the ways scientific understandings helped mariners, hydrographers, and naval planners imagine a transoceanic empire, this study presses military and naval historians to consider the role of knowledge implicit in the navy's hydrographic charts and the methods and practices used to make them. Historians have not commonly understood military and naval power through a scientific lens, preferring instead studies that focus on questions of military tactics, strategy, operations, and technology, biographies of generals and admirals, or the experience of combat.[5] Although historians have written institutional histories of naval science, histories of various exploring expeditions, and biographies of prominent naval scientists, the scholarship remains disparate, and in some instances quite old, requiring not only synthesis, but an analysis more tightly focused on the interplay among science, environment, and military power.[6]

Among other things, I show that the navy's scientific encounter with the sea reveals its intention to control the marine environment—albeit in different ways—long before notions of Mahanian sea power seemed to preface more explicitly imperial visions at the turn of the twentieth century. As historian Michael Reidy argues in his book on ocean science in eighteenth-century Great Britain, "an intimate knowledge of the ocean's tides and currents, its storms and magnetic variations conferred control that translated directly into the ability to dominate lands and cultures connected by its rim." The British, he contends, "did not claim the sea by fiat as they increasingly claimed land or demand legal sovereignty over it. Rather, the ocean was owned by those who could master its tides and waves, deal with its magnetic and atmospheric undulations, and traverse its waters unimpeded." The Pax Britannica, echoed historian G. S. Ritchie, was rooted in Britain's "long, steady policy of maintaining the freedom and the safety of the seas."[7] Something similar was true of the United States during the nineteenth century, though even at century's end, the American navy did not yet mirror the world dominance of Britain's Royal Navy. Let me be clear. During the antebellum era and in the American context, this notion of control was surely commercial and maritime rather than primarily military-strategic. The navy sought to promote the safe, speedy, and efficient navigation of the American merchant marine and whale fleet over the world's oceans. Charts and naval science more broadly sought to expand what Secretary of the Navy James K. Paulding in 1838 called an American "empire of commerce and science."[8] Here, control derived from improved navigational knowledge and the new ways that charts could capture, make sense of, and then recast the ocean environment back on the mariners and naval officers who sought to use it for trade and, by century's end, for imperial defense.

Although naval historians have tended to see the navy's gradual adoption of Mahanian notions of sea power in the 1890s as a departure from its historical roles of commerce raiding and coast defense, a history of American naval science, in fact, suggests more continuity between the mid- and late nineteenth centuries than historians usually acknowledge, even as the navy transformed at the turn of the century toward a very different modern steam and steel fleet built more expressly to extend American military power across oceans. While this certainly marked a departure from the days of the Old Sail Navy, I contend that the roots of America's oceanic empire lay in the maritime commercial expansion made possible in important ways by naval surveyors and hydrographers before the Civil War, and in a continuing faith among naval scientists that their work could further American national interests and maritime and naval power thereafter.[9] The Old Navy of the antebellum era was not built to command the seas as the New Steel Navy would more than a half century later. Yet its officer scientists nevertheless envisioned and helped to create a sea full of American masts astride favorable winds and currents or riding safely at anchor in foreign ports opened to them by new navigational understandings and, indeed, by revolutionary new ways of representing natural processes such as the movement of wind and water in two-dimensioned cartographic space. Control thus proceeded from knowledge and the belief that the chart could simultaneously portray and recast the sea in ways favorable to American maritime and, later, military-strategic interests.

If military and naval historians have generally overlooked science, historians of science have been equally remiss in examining the ways scientists in the military exercised important—even central and leading—roles in the practice and advancement of the discipline and, more importantly, in the ways science could be used to achieve military ends. The roots of this disconnect between historians of science and military historians are perhaps as deep as the historical rivalries between military and civilian practitioners of science that appear in this book. At a crucial moment in the professionalization of science, academic scientists often viewed their counterparts in the navy as novices, interlopers, and rivals for precious federal largess and public attention. Sometimes, naval scientists were simply wrong in theory and practice. They also forged close ties with a maritime community whose views of science and whose qualifications as scientific observers at sea seemed to civilian scientists questionable at best. Civilian scientists who boarded naval vessels often found themselves marginalized, their activities circumscribed by naval commanders, and their own philosophical interests dismissed in favor of the more practical questions of navigational knowledge that were so important to the naval officer corps and the maritime community it, in part, served.

Nevertheless, naval scientists played central roles in the early history of American science both in the navy and while temporarily assigned to work in the United States Coast Survey. Although most naval officers during the nineteenth century sought glory in battle and remained interested in scientific questions only insofar as they saw their ships safely to their destination or kept them from the ignominious dishonor of colliding with some ill-charted reef or sandbar, the fact remains that the U.S. Navy was one of the most important scientific institutions in nineteenth-century America. It was at the forefront of collecting environmental knowledge, organizing it, and making sense of it. Its work in practical or applied science as it related to matters of navigation was important and far-reaching. More importantly, as marine scientists became increasingly specialized, even as the scope of their inquiries broadened to include by the end of the century what we know as oceanography—the study not just of hydrography, but marine biology, geology, physics, and chemistry—naval science increasingly narrowed its focus to questions that were more specifically tactical and strategic as the navy embraced a muscular philosophy of sea power and then grappled with the challenges of defending a far-flung oceanic empire after the Spanish-American War. Historians of science, and of marine science more specifically, have tended to dismiss these later efforts as unimportant or tangential to the stories of theoretical and applied science they seek to tell. Between the Civil War and World War II, concludes one prominent historian of marine science, the navy's Hydrographic Office "took no more than a passing interest . . . in the advancement of oceanography."[10] Such claims overlook the central, if sometimes uncomfortable and meteoric, role that naval science played in the expansion of American military power.

Hydrographic science took many forms, from nautical almanacs to tide tables and sailing directions, but the chart was its most transformative. Tied as it was to the practical science of navigation at sea, hydrography was the purview of the U.S. Navy, whose jurisdiction to chart ocean waters extended, often vaguely, from the boundary of American coastal or territorial waters to cover the seven-tenths of the world awash in salt water. Charting the American coastline remained the job of the U.S. Coast Survey administered by the Treasury Department. While the civilian Coast Survey was staffed by many naval officers and the relationship between these two institutions was sometimes cooperative, it was more often competitive, and sometimes politically combative.[11] For both civilian and military hydrographers, the chart was the sine qua non of their work. Days in small open boats, exposed to the elements, countless heaving of the sounding lead, reading azimuths and other sextant angles—all this bespoke fidelity to cartographic method and precise quantification of the natural world. The finished chart was the culmination of a long, tedious process, representing the united work of many men and vessels, sometimes from different nations, and even disparate eras, galvanized in the service of navigation and competitive, but often friendly, scientific rivalry. Although this work could build on international collaboration and cooptation, the finished chart, marked with the seal of the U.S. Navy, its Hydrographic Office, or various exploring expeditions, punctuated with an American toponymy of cartographic names, or bisected with the tracks of American vessels, staked claims of precedence, national interest, and commercial and naval prerogative.

But while maps are tools of orientation and navigation whether on land or water, they are also fundamentally visual and visualizing instruments. "As mediators between an inner mental world and an outer physical world," wrote cartographic historian J. B. Harley, "maps are fundamental tools helping the human mind make sense of its universe at various scales." They are capable of taking omniscient views from above, seeing—at a glance—vast swaths of space, organizing, defining, and claiming it by precedence, cartographic seal, or toponymy—to be followed, then, by merchant and whaling fleets, squadrons of frigates and sloops, and later, fleets of battleships. So while sea charts at their most fundamental were essential for knowing how to get from one place to another, they were also powerful testaments to the ways Americans traversed an imagined oceanic empire.[12]

Historians of cartography have long understood maps as instruments of power whose boundaries and borders, colors and symbols, and numbers and tables cast a veil of precision and objectivity over the subjective process of constructing meaning, representation, and control. Following the pioneering work of cartographic historian J. B. Harley in the 1980s, notes James R. Akerman, historians "situated virtually all cartographic production within discourses of power and control." More recently, cartographic historians have focused on the practices and methods of surveyors, pointing to the ways the process of mapping, with its faith in empiricism and accuracy, informed a cartographic and scientific ethos that infused the map with transformative power. All this points to surveying as an imperial activity and maps themselves as instruments of empire. "To know is to appropriate," argues cartographic historian Christian Jacob. Maps, he theorizes, are "the transformation of real space into a figure ruled by laws of reason and abstraction, of the conquering appropriation of reality by means of its simulacrum." The atlas "offers a symbolic mastery of space," he argued. The ability of maps and charts to confer control—real or perceived—is thus a central claim in the historiography of maps and mapping. While armies and navies have long been considered as instruments of national power and empire, military historians have often overlooked or downplayed the role of their scientific and cartographic efforts in this process and, in particular, the ways mapping and charting established claims to mastery over the natural world.[13]

Nowhere are these processes more starkly drawn than at sea, where navigators barely saw with their own eyes what existed beneath the hulls of their ships, what dangers—real or imagined—lurked in the ocean's depths. Historians of cartography have examined terrestrial maps and topographical surveying as powerful symbols of imperial control. The cultural gaze of empire, they argue, is nowhere more potent than in the cartography of the nation-state or the imperial metropole. Indeed, in colonial and nineteenth-century America, historians have shown that geographical consciousness and cartographic representations were integral to the growth of the nation and the construction of national identity.[14]

Charts were likewise central to the expansion of empire, and so here I push historians of cartography both outside the borders of the United States and offshore, away from the well-known contours of the terrestrial map toward the watery edges of the hydrographic chart. Historians of cartography have not given the latter similar attention. While the historiography has focused on the duality of the map's power and its limitations as a representation of the natural world, the vast majority of this scholarship, with the exception of a number of works on the portolans of the medieval Mediterranean and the sea charts of Early Modern Europe, has focused on terrestrial maps. These historians have left largely unexamined the peculiar power and problems of cartographic method and representation in a watery world.[15] While sea charts share much in common with land maps, the two were also different in fundamental ways. At sea, the process of mapping took on many new and unique dimensions. The environment posed challenges to cartographic practice and production that exposed the chart's limitation to capture a dynamic and multidimensional space on paper. From the ocean surface to the sea floor, from the atmosphere above to the tidal zones, shifting sandbars, and reefs whose contours and whose very nature was alive with marine organisms, the ocean environment distorted all that was linear and two-dimensional. The sea chart, then, took on burdens of representation that made even the terrestrial environment seem static by contrast. Whatever its conceit of precision and comprehensiveness, often the hydrographic chart remained like a faded and shattered mirror of the environment it purported to represent—distorting and dissembling as much as it revealed. In many cases, though, the chart was all the navigator had to visualize the marine environment, which lent it authoritative power despite inherent fallacies or flaws.

A final set of historical questions draws on scholarship in environmental history, which embraces a broader view of historical agency to include both humans and the natural world as actors. Environmental historians believe that historical change happens at the intersection of human and environmental factors.[16] Historians outside this field, including many military and naval historians, remain generally skeptical about, if not opposed to, notions of environmental agency. Environmental historians, however, do not subscribe to environmental determinism. As J. R. McNeill has written, the natural world did not "strictly speaking, determine the outcomes of struggles for power, but [it] governed the probabilities of success and failure in military expeditions and settlement schemes."[17] We cannot fully understand the growth of American commercial and military power at sea without acknowledging the agency of the natural world; we cannot know the chart's power and its limitations without considering the marine environment and Americans' often profound experiential encounter with it; we cannot appreciate the particular difficulties of scientific practice at sea without reckoning with the ocean, which often belied attempts to know, control, and change it.

Perhaps no environment makes a better case for the agency of the natural world. Mariners long personified the ocean's power as King Neptune or Davey Jones or God. The sea is omnipresent in seafaring. Take, for example, the experience of Passed Midshipman William Reynolds, a young officer-surveyor in the United States Exploring Expedition whose voyage from 1838 to 1842 I examine in chapter 2. Reynolds commanded a surveying launch from which he and his men were to take precise measurements as part of the construction of hydrographic charts. He found, however, that his boat hardly constituted a firm and immobile observation platform. "We did the best we could," Reynolds admitted, referring to his survey of one of the Samoa islands in 1839. "If you can imagine yourself upon the back of a wild Horse, tearing madly over a broken and stony causeway, with precipices on either hand and the fall and roar of torrents almost stupefying your senses, you can have some idea of the critical chances of a whale boat in a sea way and of the feelings of her crew," he recorded. Feelings were laid bare, of course, but so were the survey's precision and comprehensiveness and so, by extension, the chart that emerged from it. Reynolds's nerves were "fairly fried," but he was thankful that "they had been tolerably well strung by previous experience." His boat adventure, in other words, was not exceptional. The sea's dynamism was the norm; it is everywhere in surveying and cartographic accounts, and in the journals of mariners more broadly.[18] In treating the marine environment as an actor, which, along with human contingency, influences historical change, historians can more deeply understand the process of American empire building, whose very nature American surveyors and chartmakers struggled to capture visually and to conquer cartographically.

Natural agency presses historians to consider the possibilities but also the limits of human intention and will in the natural world. Whatever the prevailing biblical, folkloric, scientific, or technological ethos of the day, the environment is not easily conquered or harnessed for human use. Hydrographic surveying and charting is, by nature, always unfinished, perpetually being undone by the marine environment itself. Indeed, an environmental analysis of American naval and maritime history suggests that the marine environment has been a kind of natural enemy that not only undermined American efforts, but threatened life and property, fundamentally influenced the structure, technology, war making, and culture of the navy, and sometimes proved an important factor in the outcome of naval operations in war and peace. In some cases, it proved every bit as menacing—if not more so—than the enemy. During the Spanish-American War of 1898, for example, naval commanders on the inshore blockade of Cuba in chorus lamented the difficulties of navigation and the shortcomings of their charts. The war, observed the navy's chief of the Bureau of Equipment in 1903, "would have assumed a totally different aspect" had the navy "possessed the charts which have been constructed from surveys made since the Spanish War."[19] While historians of the war have tended to emphasize the navy's dominance over its Spanish enemy in what many then and since deemed a splendid little war, the bureau chief's remarks suggest that a more nuanced analysis focused on the marine environment and the navy's cartographic knowledge could shed new light on long-held interpretations about the ways that war unfolded. American sea power, broadly defined, was never far removed from knowledge of the sea itself. It derived, in important ways, from control or perceived mastery of the natural world.

A growing body of scholarship in war and environment has brought attention to the centrality of the natural world to armed conflict and to military institutions in general. In the last decade, historians have begun to transcend studies of terrain and geography in military operations to examine the environment as a natural enemy or ally whose agency can, among other factors, influence the outcome of conflict. The side that changed or used nature most effectively was often victorious, as historians Lisa Brady and Megan Kate Nelson have shown in environmental histories of the American Civil War. Such advantages often derived from cartographic knowledge on land and at sea. Moreover, these historians have also shown that conquest over one's enemy is often tied to related notions of conquest over nature rooted in the engineering and scientific specializations that in many instances defined the military profession from the late eighteenth to the mid-nineteenth centuries in Europe and the United States.[20]

These studies press historians to see the environment not just in material terms but also in imagined ones. For mariners and hydrographers of the nineteenth-century ocean, the marine environment existed on two planes. The first was the material world of water, rocks, reefs, and sand-bars of the littoral or the deep ocean's muddy ooze. The second existed as a set of emotions and ideas associated with often profound and meaningful experiential encounters with the sea. Historians must acknowledge cultural attitudes toward nature, or the ways in which humans conceived of the environment and their place in it. The sea is a material space to be sure, but it cannot and should not be separated from the sea as a social construct. "The sea remains . . . a space constructed amidst competing interests and priorities," argues the social scientist Philip Steinberg, "and it will continue to be transformed amidst social change." Antebellum naval scientists, for example, often perceived the sea as a sort of watery wilderness similar to, but also fundamentally different from, the notions of land wilderness that environmental historians have long examined as central to the expansion of American settlement and national power across the continent.[21] While American mariners navigated the sea according to their own sense of folkloric and scientific meanings, naval scientists interpreted the sea as a wilderness in need of orientation and ripe for the kind of order and control the surveyor, the mapmaker, and the chart itself supposedly imposed. Control or conquest of human and natural forces was therefore the result of battlefield victories and diplomatic treaties, but it also existed as a state of mind that envisioned the spread of American influence over the ocean not by felling trees and fencing land, but by understanding the marine environment and representing it in ways that harnessed it in the service of larger national aims. In this way, hydrography mediated the relationship between humans and the material and imagined environment of the ocean. At least until the mid-nineteenth century, the ocean environment could not be altered in discernible ways as land could be "improved." It could, however, be recast and reenvisioned in the mind. In this process, the chart was indispensable.[22]

More than a decade ago, the maritime and environmental historian W. Jeffery Bolster surveyed the field of environmental history, calling attention to the need to historicize the sea. He might have been speaking to so many more historians in fields outside environmental history whose focus remained land bound. For Bolster, the sea lacked a historiography comparable to land. He argued that historians long perceived it as a space "outside of time and beyond the pale of history."[23] Since Bolster wrote that line, historians have gotten their feet wet, largely, it seems, because the ocean itself has lately become the focus of so much attention over the question of human-made environmental changes of a sort that, in part, protected Tubbataha Reef from destruction by everything from ship hulls to overfishing.[24] Even as this growing body of scholarship has begun to historicize the sea, giving context to our shared relationship with this most important environment, it has also begun to revise our broader narratives of war, politics, class, and race.[25] Attention to humans' relationship with saltwater, in other words, is not only important; it is essential for a fuller and deeper understanding of the past. *To Master the Boundless Sea* follows these works and—I hope—moves in new directions not in order to discount or overturn older interpretations of American power and imperialism, but to transcend them and, in doing so, to find in the sea a thread that can tie together the often disparate fields of military, naval, and maritime history with environmental history and the history of science and cartography. Set on, in, and under water, its main characters are the marine environment itself and the naval scientists who hoped and struggled to understand, represent, and thus control it. In all this, the sea matters. In the ocean, that is, we can better understand the ebb and flow of power and the pitch and roll of history.

# 1 Wilderness of Waters

Nothing is lost on him that sees
With an eye that feeling gave—
For him there's a story in every breeze,
A picture in every wave.

—A Fore-Top-Man, *Life in a Man-of-War; or, Scenes in "Old Ironsides"
during Her Cruise in the Pacific*, 1841

The mariner Robert Weir closed his journal on April 12, 1858, with a common notation. "The water here has a very light blue appearance," he wrote as his ship, the *Clara Bell*, pitched in the swells of the South Atlantic. Weir and the *Clara Bell* were nearing the end of a three-year whaling voyage to the Pacific Ocean, during which, to judge from his journal, he had spent many hours reflecting on the waters that surrounded his cramped wooden world. Weir, however, followed his initial observation with a more philosophical thought. The sea's appearance, he added, "by Maury's Theory denotes the extreme depth." He was referring to Lieutenant Matthew Fontaine Maury, superintendent of the United States Naval Observatory and Hydrographical Office in Washington, D.C., who, as Weir hunted whales, was preparing edition upon revised edition of his *Wind and Current Charts*. These he amassed with the help of shipmasters and navigators who sent him their observations about winds, currents, whales, meteorology, and many other natural phenomena of the sea. With these charts, Maury was revolutionizing voyages under sail by harnessing the power of winds and currents in the service of American maritime enterprise. Weir had presumably read something of Maury, and so he headed aft to consult a chart. "Soundings are marked on the chart hereabouts from 3 to 4000 feet deep," he observed before concluding with the remark, "Could we see the waters cleared at such a depth—what awful and mysterious wonders would we behold—."[1]

In some ways, Maury's charts promised to do exactly that—to reveal to the mariner a world beneath the waves that had long been shrouded in mystery and conjecture, a world that mariners like Weir sought to know through a mix of folkloric and scientific understandings. In Weir's journal entry, there is a sense of growing faith in an emergent marine science. There is perhaps also—in the last dash of his pen stroke—a respect for the ocean's secrets and an appreciation for the chart's limits in opening it to the mariner's eye. American naval science emerged in the early nineteenth century just as the nation's maritime commerce was expanding into new waters and tapping new markets in the Pacific and East Asia. The navy established its Depot of Charts and Instruments in 1830, slowly expanding over the next three decades to provide charts and sailing directions to a growing merchant marine and whaling fleet. At sea, shipmasters and navigators adopted naval scientists' practical and theoretical work, which promised safer, faster, and more inexpensive voyages, but many also retained their own folkloric understandings of the marine environments' workings rooted deeply in the experience of the maritime community. It was precisely these vernacular maritime understandings that American naval scientists sought to dispel, displacing what they deemed to be an ocean wilderness with the ordered empiricism of hydrographic science. In doing so, they constructed a new vision of American commercial empire on the ocean.

Seafaring in Weir's day was dangerous. Mortality rates for mariners in the age of sail are elusive, but maritime and naval historians generally agree that the number of Americans who died at sea in the nineteenth century is significant. Historian Daniel Vickers, who studied the port of Salem, Massachusetts, concluded that in the eighteenth century, the rate of widowhood was twice that of the primarily agricultural towns that surrounded the port. By the nineteenth century, he contends, "a sailor's prospect of dying within a decade at sea was . . . about 30 percent." Naval historian Christopher McKee notes that American naval officers were three times more likely to die "lost at sea" than in battle in the early nineteenth century. "An often overlooked aspect of life and death in the pre-1815 navy," he observes, "is that a substantial number of its vessels sank with all or part of their companies. For the . . . naval officer the great occupational hazard was the sea itself." Seafaring constituted some of the most dangerous work in the Early Republic. Of course, mariners died from many causes such as disease or work-related accidents, which were not, at least directly, attributable to the dangers of the marine environment. Nevertheless, ships sank or simply disappeared without a trace, the victims of violent storms, navigational error, uncharted reefs, or rocky lee shores that made short work of wooden hulls.[2]

The experience of seafaring in the early nineteenth century rested fundamentally in a sustained encounter with the marine environment. One's very survival was often at stake, but more important, in an Age of Sail when harnessing winds and currents was vital to the daily operations of the ship, and thus the routines of its mariner-laborers, the majority of whom were employed setting, reefing, and taking-in sail, the sea was a constant presence, continuously observed, pondered, and divined. Aboard the bark *Magdala* bound from New York to San Francisco in 1849, passenger Humphrey Hill observed the captain to be "very passionate and very disagreeable . . . except when we have a fair wind; he is then somewhat more agreeable." For the *Magdala*'s captain, everything depended on the wind. While the experienced shipmaster often had general ideas of the wind's direction in any given place, winds were so variable, dependent on circumstantial conditions and seasonal changes, that even the most seasoned captain had but a vague idea of what he might encounter during the course of a voyage. He did not so much use the natural world as respond to its variability, hoping that his own reactionary decisions hastened the voyage to a successful conclusion.[3]

More broadly, as mariners looked to the prospects of a long deep sea voyage, it was the ocean itself that affected them most profoundly both in material and imagined ways. Mariners' journals are replete with the sea sickness of the green—or inexperienced—hand, usually followed upon regaining his sea legs by feelings of reverie and awe at the sea's vastness and its power—a phenomenon Richard Henry Dana called "the witchery in the sea" in *Two Years before the Mast*. Often, though, the mariner's enthusiasm for his surroundings soon gave way to boredom and monotony. Despite Robert Weir's constant reminder to see God's work in every wave, he nevertheless lamented, "no sail or land . . . to break the even line of the horizon; all around is that same dreary and desolate waste of waters that we have so long and often beheld." An unbroken expanse from horizon to horizon, the sea's vastness could be disorienting. In Weir's case, it became simply ponderous. In 1849, as the whaleman Nelson Cole Haley looked toward a voyage that might last three or four years, he confided his resignation to his journal. "Stormy gales, hurricanes, lee shores, and whales jaws, and flukes. For what?" he grumbled, "not for money!" Mariners such as Haley and Weir inhabited a world interpreted through the movement of wind and water. The sea conjured moments of awe and inspiration, hair-raising danger, and, most often, a sense of omnipresent and oppressive monotony.[4]

Mariners understood the natural world most intimately through labor. As maritime historian Marcus Rediker has written, the "outlook" of sailors was "fundamentally shaped by the nature and setting of their work . . . based upon an essentially materialistic view of nature, a desire to make an omnipotent nature seem orderly and comprehensible, and a need to entrust to each other their prospects for survival." Aboard a sailing vessel, changes in wind speed and direction, the unexpected discoloration of water under the bow indicating dangerous shoal water, or the advent of storms elicited sharp orders from the captain, the master, or his lieutenants or mates that sent men scrambling up shrouds and scurrying along yardarms. On deck, others hauled on lines, singing halyard chanteys or maritime work songs that spoke to the centrality of the natural world to life and work aboard ship: "We're bound away around Cape Horn/Where you'll wish to the Lord that you'd never been born/To me rollickin' Randy Dandy-O/Around Cape Horn we all must go/Way off 'round Cape Stiff through the cold rain and snow/To me rollickin' Randy Dandy-O." Nature set the course, speed, and rhythm of life in the Age of Sail, and so it is not surprising that in the early nineteenth century, mariners sought to understand the marine environment through their own experiences, through a shared folkloric knowledge, and, increasingly, through the tools of navigational science—the chart, the sextant, the nautical almanac, and various coast pilots and books of sailing directions that oriented the mariner's place from the deep ocean to the shallow littoral.[5]

By the early nineteenth century, this desire to make nature seem orderly and comprehensible, as Rediker puts it, formed a deep body of folkloric knowledge passed down from master to mate and old salt to green hand within the often tightly knit maritime communities of the Early Republic and the even more tightly woven—though rigidly hierarchical—fabric of the ship's officers and crew. Sailors often whistled in imitation of a stiff breeze through the ship's rigging, for example, in hopes of conjuring the sailing vessel's only motive power. Many swore against the killing of albatrosses as Coleridge's "Rhyme of the Ancient Mariner" later taught so many schoolchildren. Folkloric knowledge often manifested in sayings, songs, rhymes, and poetry that constituted this largely oral tradition. "Mackerel skies and mares' tails," referring to evocative cloud formations "make tall ships carry low sails. If clouds look as if scratched by a hen," the rhyme continued, "get ready to reef your topsails then." Sayings such as "if the sun sets clear as a bell, it's going to blow sure as hell," helped the mariner divine future weather conditions. For many years, observed the folklorist Horace Beck, navigators making a transatlantic crossing would follow the maxim to sail southward " 'till the butter melts," at which point they would find the trade winds. Here, there was little need for sextants and nautical almanacs. So-called dead reckoning, a rough estimation of the distance traveled by speed and noon sightings by sextant, characterized the daily practice of many a ship's captain, sailing master, or navigator in the early nineteenth century.[6]

In the United States, naval science emerged within a broader maritime world in part beholden to these vernacular understandings. Such beliefs were meaningful; they were not easily cast aside for new and different ideas or systems of knowledge, and often they proved effective enough in getting vessels to their destinations. "Superstition," the American writer James Fenimore Cooper observed, "is a quality that seems indigenous to the ocean." It was certainly indigenous to the men who struck out over the horizon, often with little more than a basic understanding of spherical trigonometry, but with hard-won knowledge gleaned from previous voyages. In his book *Two Years before the Mast*, published in 1840, Richard Henry Dana gave a sense of what the convergence of folklore and scientific rationalism actually looked like aboard a merchant vessel at midcentury. He recounted a conversation with the ship's cook about a Scandinavian shipmate whom the cook thought to have special powers. The cook, Dana wrote, "was fully possessed with the notion that Fins are wizards, and especially have power over winds and storms. I tried to reason with him," the Harvard-educated Dana wrote, "but he had the best of all arguments, that from experience at hand, and was not to be moved." To Dana's suspicions, the cook replied: " 'Oh . . . go 'way! You think, 'cause you been to college, you know better than anybody. You know better than them as 'as seen it with their own eyes. You wait till you've been to sea as long as I have, and you'll know." While the masters and mates whose vessels naval scientists and their charts aided might also have looked askance at the cook's exchange with Dana, it nevertheless suggests the sort of suspicion with which outsiders, and particularly those privileged with elite, academic backgrounds, were held by mariners whose own observations seemed the most viable source and system of knowledge. As historian of science D. Graham Burnett has observed, "laboring folk . . . had their own ideas about how to organize the natural world, ideas derived from labor, craft, scripture, and experience." The challenge for naval hydrographers, whose work served and relied not only on the navy but also on the larger American maritime world, was to replace these folkloric meanings with new scientific understandings and cartographic representations based not in vernacular experience, but in Enlightenment empiricism and rationalism.[7]

In this transformation of navigational knowledge, we can see the expansion of American commercial empire at sea as the American merchant marine and whaling fleet entered a period of extraordinary growth between

1815 and 1861. There is little doubt, now, thanks to the work of Amy S. Greenberg, Paul Lyons, Dane Morrison, and Brian Rouleau, among others, that this era of maritime commercial expansion was imperial in its causes, conduct, and outcomes. Herman Melville's Nantucketers, "issuing from their ant-hill in the sea" having "overrun and conquered the watery world like so many Alexanders" was not all literary hyperbole. "Let America add Mexico to Texas, and pile Cuba upon Canada," he wrote in *Moby-Dick* two years after the Treaty of Guadalupe-Hidalgo, "two thirds of this terraqueous globe are the Nantucketer's." At once characterized by resource extraction, modest but important diplomatic and territorial claims such as the Guano Act of 1856, violence and economic exploitation against indigenous people, and scientific and nationalistic rivalry with European powers, American mariners formed the vanguard of a vigorous oceanic empire, particularly in the Pacific, where trade in spices, sandalwood, the sea slug known as bêche-de-mer, and, of course, whaling, made fortunes for merchants and shipowners, and made cities such as Salem and New Bedford, Massachusetts, commercial centers.[8]

In this era, the mariner emerged in American literature and, in some cases, quite consciously to himself as a sort of frontiersman gone to sea, an archetype that, in large measure, derived from his encounter with the marine environment, its uncharted dangers, and its epistemological mysteries. The American writer James Fenimore Cooper, long studied by environmental and cultural historians for his vivid portrayals of the early American frontier in his Leatherstocking Tales was also a prodigious chronicler of the sea. Cooper had himself served in the merchant marine and had written an early history of the American navy. In sea novels such as *The Pilot*, *The Red Rover*, and others, Cooper established an American frontier at sea while making analogies in his other work to the connection between the American land and sea frontiers. He wrote of the prairie, for example, as resembling the vast, uniform, and rolling spaces of the open ocean. Cooper's mariner-hero, Long Tom Coffin, was a sort of salty Natty Bumppo. By the 1850s, Herman Melville was writing about the whaleship as "the pioneer in ferreting out the remotest and least known parts of the earth. She has explored seas and archipelagos which had no chart, where no Cook or Vancouver had ever sailed." Together, Cooper, Melville, and others such as Edgar Allan Poe, constructed the sea as a space of American empire, of an emergent national identity, and a place where American mariners did battle with a hostile nature. Sometimes they triumphed; sometimes few lived to tell the tale, but it is no coincidence, as I argue in chapters 2 and 3, that the sea fiction of Cooper, Melville, and Poe inspired and drew upon the work of naval scientists in framing literary geographies of an emergent oceanic empire.[9]

In constructing this sea frontier, American writers were not just waxing Romantic about some idealized ocean; rather, their work reflected a very real sense among mariners that they were transcending new boundaries of space, knowledge, and exploration in extending the American flag to supposedly savage places, peoples, and environments. William Abbe, a mariner aboard the whaleship *Atkins Adams* during a voyage in the 1850s, for example, interpreted whalers like himself as "a fearless seaman, penetrating among the tumbling mountains of ice in the blustering outrageous Northern Seas, or under the frozen Serpent of the South, or he vexes with his hull keel the milder waters of the Indian or Pacific Oceans." Abbe's journal is striking for the ways he frames American oceanic frontiering as a struggle between mariners and the ocean environment: "Geography and empire honor him among their best contributors, but it is no love of fame, no ambition of honor, no desire to assist science or increase knowledge—to make discoveries or carry his country's flag to unknown shores—that makes him a hardy navigator and a bold discoverer. It is the pursuit of oil and the chase of his tremendous prey. It is the interest of owners—of himself and crew—not the interest of mankind that urge his keel among unknown seas and amid unusual dangers." Clearly, Abbe articulated a vision of American commercial empire that was not scientific or even territorial, but rather premised on commerce.[10]

Maritime trade was central to the growth of the young nation and the construction of national identity. "America faced East," argues historian Dane Morrison, seeking to reorient narratives of national identity and westward continental expansion. "The fundamental experience during the formative years of the new nation was lived on the waters that led to China, India, and Java," he concludes. In 1838, on the eve of the navy's first voyage of exploration to the South Pacific, the nation's maritime trade was worth $223 million, nearly 90 percent of which was carried in American holds. In the late 1840s, more than seven hundred of the nine hundred vessels hunting whales in the world's oceans flew the American flag. At its height, seventy thousand Americans were involved in some aspect of the whaling industry, representing a capital investment of $70 million. In antebellum America, Manifest Destiny also meant maritime destiny, to paraphrase Brian Rouleau. "Saltwater itself," he writes, "was spoken of as a field for national domination." Together, Morrison, Rouleau, and others press historians to think of American imperial expansion not just in terms of the westward movement of people over land, but in a multidirectional web of maritime voyaging that pushed the boundaries of American commercial and cultural influence around the world. "Nineteenth-century Americans envisioned national empire extending into the Atlantic and Pacific," Rouleau argues, "not halting at their shores." Like so many aspects of American society, Alexis de Tocqueville seemed to capture the reckless expansionist spirit of maritime enterprise when he wrote in *Democracy in America* that "the Americans are often shipwrecked, but no trader crosses the seas so rapidly. Americans affect a sort of heroism in their manner of trading. They are born to rule the seas, as the Romans were to conquer the world."[11]

As American mariners pressed their bows farther from the nation's shores in search of whales, new markets, and lucrative trade in the period after the War of 1812, the United States Navy expanded and followed them. The navy "played a positive and expansive role in the nation's burgeoning overseas commerce and in its emerging American maritime empire," writes naval historian John H. Schroeder. Commodore Matthew C. Perry, the naval diplomat who engaged Japan and set the stage for the Treaty of Kanagawa, foresaw a future in which "the people of America will, in some form or other, extend their dominion and their power, until they shall have brought within their mighty embrace the islands of the great Pacific, and placed the Saxon race upon the eastern shores of Asia." By 1850, the navy had established five permanent squadrons stationed across the world at important points for the protection and promotion of maritime trade. The "objects of our government in keeping a naval force in these seas," wrote the commanding officer of the sloop-of-war *St. Mary's*, are "to afford aid and protection to our commerce and to look after the interests of our citizens generally, where engaged in their lawful pursuits." Sometimes such aid turned violent. Commodore John Downes, while in command of the frigate *Potomac* in 1832, burned Kuala Batoo, Sumatra, in reprisal for the pillaging of an American merchant vessel; USS *Cyane* leveled Greytown, Nicaragua, in support of Cornelius Vanderbilt's Accessory Transit Company in 1854. These are just two of many incidents that are illustrative of the way the navy used force to further the interests of American commerce between the War of 1812 and the Civil War. The navy was not yet a force built to rule the seas, to borrow de Tocqueville's phrase. Nevertheless, naval historian Kenneth Hagan argues that naval policy after the War of 1812 was "well calculated to meet the needs of a secure continental power with extensive maritime interests."[12]

One pressing need among merchant and whaling captains and naval commanders was for charts of unknown or little-known waters. Even at midcentury, much of the world's oceans remained uncharted. Knowledge was better along some coastlines, particularly those of Europe and the eastern United States, where passage in and out of port cities privileged navigational accuracy in busy waters, but even along the American coast, the hydrography was complex and often only half-known, commonly requiring the continual heaving of the lead line or sounding wire to literally feel one's way onto and along the shallows of the Continental Shelf. "The life of the most experienced [mariner] is more endangered when he approaches the coast, than when exposed to the tempests which agitate the mid-ocean," remarked the well-known publisher of navigational texts Edmund March Blunt, whose *American Coast Pilot* became, by the first decades of the nineteenth century, the standard guide for navigators seeking entry to American ports. The deep ocean remained virtually uncharted since navigators were most concerned with the shallows where their hulls might meet reefs, rocks, and shoals to disastrous consequence for life and property. Yet even on the deep sea, mariners feared *vigias*, or reputed navigational dangers, such as rocks or banks that rose, often suddenly, from the deep sea and whose location, and even whose existence, was vaguely known, reported by some passing ship, but not thoroughly investigated. "We feared for some time over water which was very much discoloured from which circumstance we supposed ourselves on the dangerous bank which is so variously described in the different charts and have hauled our course more to the southward to avoid all the dangers, whether real or imaginary," wrote a sailor aboard the schooner *Weymouth* bound for Java in 1823. Two days later, he recorded in his journal once more, "These dangers are scarcely to be depended on, there are several laid down under several different names as seen by so many different navigators." The charts on which the American maritime world relied were full of blank spaces and half-truths, and so the navigator was inclined to view his chart with some authority, but also with measured suspicion for what it might or might not reveal to him about the nature of the marine environment.[13]

Shipwreck narratives became a popular form of American writing in the early nineteenth century, conjuring images of the natural world—particularly uncharted reefs—that served both to dramatize the adventurous voyages of these mariner-frontiersmen and to call public attention to the need for more accurate hydrographic knowledge of waters newly important to American trade. By the 1830s, for example, the Fiji Islands of the

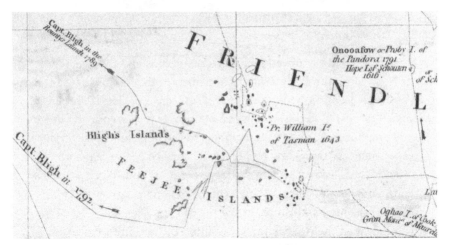

Detail from British chartmaker Aaron Arrowsmith's *Chart of the Pacific*, 1798, showing the tracks of Bligh and Tasman with the half-formed coasts of the Fiji Group. This chart went through a number of updates by the 1830s, but the Fijis remained in an incomplete state of navigational knowledge. Courtesy of the Library of Congress, Geography and Map Division.

Southwest Pacific had entered the American consciousness as a destination of many unknown and mysterious dangers. Traders out of Salem, Massachusetts, began arriving in the 1790s to tap a lucrative trade in sandalwood. When the groves were exhausted, they turned to harvesting bêche-de-mer, a reef-dwelling sea slug prized in the China trade as an aphrodisiac. Fiji was an infamous graveyard, a natural environment so mazelike in its uncharted reefs that even the famed British explorer James Cook had declined to explore them in the 1770s. By the early nineteenth century, the newspapers of maritime New England were littered with reports of shipwrecks and the cannibal customs of the Fijian people. As I explore in chapter 2, uncharted reefs and rumors of cannibalism went hand in hand as two spectral sides of environments Americans hoped to control through knowledge and, if necessary, through violence.[14]

Americans increasingly found waters like Fiji's alluring for the value of their natural resources and, ironically, for the horrifying habits of its people. Sailing among the Fijis "was rendered extremely dangerous and difficult," wrote William Endicott, "by reason of the imperfection of our chart, and the islands being improperly surveyed." Endicott was third mate of the ship *Glide*, also out of Salem when it wrecked on an uncharted reef in 1829, and along with his shipmate James Oliver, published an account of the voyage.

These contributed to Fiji's cultural and environmental notoriety. Oliver recounted the moment of impact:

> Thus cautiously we moved along this dangerous archipelago, with the helm alternately at starboard and port, as the lookouts gave the word. The occasion was one of the most intense anxiety. Every man was at his post, and the stillness on board was broken only by the loud and hurried orders of the officers to the helmsman. The crisis was at hand. Within a few feet from the surface of the sea, and directly ahead, was discovered a large coral rock. Instantly vociferated the officer aloft, —hard down the helm! —hard down the helm! Shouted the man amidships, at the top of his voice—hard down the helm! Replied the steersman, as he suited the action to the word. The next instant, the *Glide* struck with a tremendous crash.

Endicott and Oliver's shipwreck narratives and subsequent tales of cannibalism are characteristic of a genre of South Sea tragedies begun by the American John Ledyard's well-known account of James Cook's death by Sandwich Islanders in 1779, as well as Captain David Porter's *Journal of a Cruise Made to the Pacific Ocean*, published in 1815, chronicling, among other things, the American frigate *Essex*'s time in the Marquesas Islands, which, like Fiji, were supposed to be inhabited by cannibals. Together, this body of literature constructed an image of the Pacific and its indigenous people as both dangerous and curiously exotic. These were "savage coasts," Oliver wrote, as if the reefs offshore somehow took on the specter of Fijian cannibals themselves.[15]

Indeed, the uncharted reefs that spelled the end of many vessels and captured the attention of the American maritime community and a broader reading public represented a peculiar aspect of the marine environment that Americans sought to control both materially and imaginatively. Reefs were the focus of widely held dread because the cragged calciferous shells on which myriad marine organisms slowly built them broke to pieces any ship unfortunate to encounter one. Many reefs had garnered their own particular notoriety. As Mary Wallis, the wife of a captain from Salem, Massachusetts, observed in 1851, "nearly every island and reef that we pass has its incident." When the *Salem Gazette* reported in 1835 the loss of the whaleship *Albion* by a "too intimate acquaintance" with a Fijian reef, it suggested that such environments were more than just impassive or objective navigational hazards. Rather, they had become infused with meanings rooted in the encounter between humans and the natural world. Reefs, however, were environments of many different meanings. Neither water nor land, they presented the avenues on which Fijians moved from shore to ship, seeking to trade or—perhaps—to pillage. The reefs were also the homes of the sea slug the Americans prized and, consequently, the site where Fijians, in the employ of American traders, fished for them. The reefs themselves were thus complex spaces of American commercial empire in places such as the Fiji Islands, at once fraught with danger, but also with commercial and imperial possibilities that more accurate hydrographic charts and more comprehensive navigational knowledge could exploit.[16]

Navigational dangers real and imagined dominate the popular sea narratives that captured the public's attention and served both as warnings and guides to those who followed the writer's wake. Literary scholar Hester Blum has examined these texts written by American sailors during the Early Republic and antebellum eras, arguing that despite stereotypes of illiteracy—among many other supposed vices—mariners were, in fact, literate consumers and producers of books. Shipboard labor, she contends, became "the basis for both applied and imaginative knowledge," which "allowed [mariners] to 'read' the sea as a text." Among the usual navigational treatises found in many naval and merchant vessels' libraries, mariners could turn to accounts written by fellow sailors. In the *Narrative of the Shipwreck of the Sophia*, for example, the sailor Charles Cochelet wrote for those who "in future receive instructions to avoid a shore, which has already proved fatal to so many human beings." In 1818, James Riley wrote an account of his shipwreck and captivity in North Africa with the intention that "mariners, particularly, being thus apprised, will guard against the constant currents which have caused such frequent and dreadful disasters as death, slavery, and other almost incredible sufferings." Riley's purpose was "to warn the navigator of the danger he is in when his vessel is acted upon by these currents, where no calculation can be depended upon." In an era before institutionalized navigational science in the United States, mariners relied, in part, on a growing body of sea writing that spun good yarns for an American readership increasingly interested in such stories, but that also served to contextualize their own encounter with the marine environment and to leave a testament to an experience that might serve as a warning or a guide to fellow navigators. Much more than just sea stories framed in the narrative arc of the voyage, these works were literary cartographies or a kind of textual map that constructed environmental knowledge in the sailor's imagination. Such appeals were not lost on naval scientists, the most notable of whom, as we will see in chapters 2 and 3, produced lengthy and sometimes exquisitely written narratives of their own meant to supplement and narrate the charts that accompanied them. Cartography and literature, text and map, word and image were inextricably linked in this expanding American maritime world.[17]

Among the most important of this genre is David Porter's *Journal of a Cruise Made to the Pacific Ocean*, a two-volume narrative recounting Porter's commerce warfare against British shipping in the Pacific during the War of 1812, replete with page-turning stories of battle with enemies both human and natural. When James Fenimore Cooper wrote about this voyage in his *History of the Navy of the United States of America*, published in 1839, he not only chronicled the commodore's sea battles but also noted that Porter had "but one chart of the ocean in the ship and that," he added, "was very small and imperfect." Clearly for Cooper, whose literature fused land and sea frontiers and who saw American identity constructed as much in watery battle with a hostile nature as in the wilderness of the North American frontier, Porter's wont of charts mattered because it suggested something of the feats that America's new naval heroes could attain against adverse odds. Like so many American navigators, Porter seemed to embody in his own account, as well as in Cooper's estimation of him, the sort of reckless American mariner-adventurer that de Tocqueville had prophesied was born to rule the sea.[18]

Porter himself framed his voyage to the Pacific as a navigational exploit conceived to overcome nature, burnish his reputation as a great and fearless navigator, and dismiss European predecessors in exploration, even as he took the fight to the British whale fishery. "Many readers will perhaps find some of my nautical remarks dry and uninteresting," he admitted with a sort of self-abnegation that belied his own sense of literary craftsmanship. He knew well enough how to spin a good yarn. "Navigators will view them differently," he continued, "and will not esteem them the less for not having had the ornamental touches of a fine writer." Porter seemed to be speaking to fellow sailors, not in the language of the armchair navigator, imagining an ocean wilderness, but as one who had actually experienced it and—perhaps—conquered it even as it had tested the physical endurance of himself, his crew, and his ship, all common themes in voyaging narratives of the early and mid-nineteenth century.[19]

In part, then, Porter's narrative might be read as a navigational text broadly defined, in which—like Cooper who later wrote of his exploits—he sought to conquer a marine environment and stake an American claim to Pacific waters through navigational prowess. These conquests of nature—

articulations of an emergent ocean empire—were not primarily diplomatic or military in a traditional sense. They were cultural, framed in the medium of voyage narratives and navigational texts written for popular audiences as much as for fellow mariners. Consider the following excerpt from Porter's rounding of Cape Horn:

> I shall take this opportunity of offering some hints to those, who may succeed me in attempting the passage around Cape Horn; and this I feel myself the more authorized to do, as we have effected it in, perhaps, a shorter time, with less damage, and laboring under more disadvantages, than any others who ever attempted it; and that too by struggling, at an unfavorable season of the year, against a constant succession of obstinate and violent gales of wind. And I am the more strongly induced to offer these hints, conceiving it to be of the utmost importance to navigation, to give any information, derived from experience, which may tend to enable navigators to overcome the obstacles which nature seems designedly to have placed, to deter mankind from all attempts to penetrate from the Atlantic to the Pacific ocean; and, as various opinions have been given on the subject, my advice may differ from that of others in several points: but as my measures have proved successful in the end, and as it is not founded on mere conjecture and hypothesis, it is to be presumed, that it may deserve the attention of seamen, for whom alone it is intended.

Here was valuable firsthand knowledge of a treacherous ocean passage, but one that also made for stirring and evocative reading as Porter set himself up against a recalcitrant nature whose "designs" he had faced head on and ultimately overcome.[20]

Porter also situated himself relative to European explorers, revising their accounts and dismissing what he deemed inaccurate, misleading, or erroneous information with his own empiricism. He invoked "the most celebrated navigators," and "the celebrated Cook" by name, quite obviously intending to add his name to the roll. He went on, however, to refer to the "erroneous expectations, which the opinion of [the French explorer] La Perouse is unhappily calculated to lead [mariners] into, and, perhaps, has proved fatal to many ships." Here, he was both invoking Cook in order to join his fabled company while dismissing La Perouse to elevate himself and his nation in the annals of voyaging, discovery, and exploration. Even as he waged commerce warfare against the British, he took Europeans to literary war in the pages of his narrative, asserting a sort of cultural independence from the continent that was a central impulse of American scientific activities in the early nineteenth century. Porter's setting was the ocean. His common foe was the ocean environment. In many ways, such textual accounts of navigation and the marine environment would mirror the way sea charts staked national claims to waters whose dynamism otherwise flouted imperial pretensions. Texts claimed precedence, told stories, and framed oceanic environments in ways more evocative and explicit, but also remarkably similar to an emerging cartography whose own implicit narratives and meanings became central to imagining empire in oceanic terms.[21]

It is no coincidence that the institutionalization of hydrography and navigational science more generally coincided with the expansion of the American maritime world. The navy established the Depot of Charts and Instruments under the command of Lieutenant Louis M. Goldsborough in 1830. It had modest beginnings. During its first years, the depot in Washington represented little more than a storehouse for the navy's nautical instruments such as chronometers, theodolites, sextants, and hydrographic charts. Its origins reflected, in particular, the pressing needs of both the navy and the American maritime community for more and better charts and, more broadly, the emerging place of scientific work within the navy's ranks. There was as yet no pretense to imperial projects, and even those such as the voyage of the United States Exploring Expedition, 1838 to 1842, were only partly directed by the depot, remaining largely under the direct supervision of the secretary of the navy. Nevertheless, over the next eighty years, the cumulative effect of the depot and its successors' work and that of the surveyors and chartmakers under its direction would play a significant role in propelling the United States onto the world stage first as a commercial power and later as a formidable sea power whose oceanic empire by century's end stretched from Puerto Rico to the Philippines.

By 1844, the depot had grown into the Naval Observatory and Hydrographical Office under the command of Lieutenant Matthew Fontaine Maury, an institution whose role in the expansion of American overseas commerce and the collection and distribution of scientific, navigational, and cartographic knowledge assumed real transformative power and international reputation. There is no question that these efforts significantly aided the expansion of American antebellum maritime commerce. As naval historian John Schroeder has written, "the mass of valuable scientific, geographic, and commercial information produced by . . . expeditions" under Maury's supervision at the observatory, "helped to further stimulate the expansionist imagination of a nation in the process of creating a vast overseas commercial empire." Maury championed both advances in astronomic and hydrographic research, though his critics—many of whom were astronomers or who looked covetously at a naval officer overseeing a world-class observatory—impugned him for favoring the latter. From the 1840s through the turn of the century, the observatory and, after 1866, the Hydrographic Office continued this work and embarked on efforts to study and chart the depths, winds, currents, and other natural phenomena of the sea. Mostly, these efforts produced traditional hydrographic charts with depth soundings, tide tables with the imprimatur of the office and the navy to lend it authority. Sometimes—as was the case with Maury's work—the charts constructed at the observatory were new, revolutionary forms that sought to capture and convey the complex movements of wind and water. By the 1850s, it more directly organized exploring expeditions from South America to the Arctic, and, as the navy gradually embraced a more muscular vision of sea power and territorial empire in the 1890s, the Hydrographic Office took up the task of providing accurate navigational knowledge of waters newly strategic to American interests overseas.[22]

By that time, modern marine cartography had existed for more than five hundred years, first centered around the seaborne trading cities of Venice and Genoa in the Medieval Mediterranean. There, portolans such as the *Carta Pisana* set Mediterranean shipmasters on their course along rhumb lines, or compass headings, with impressive precision, thereby setting the patterns of maritime trade that knit together the continents of Europe and Africa with the Middle East. Portolans framed an emergent Mediterranean World. The origins of modern cartography and navigational science were intimately linked to seaborne trading, and so it is no surprise that the rise of mercantile empires connecting European metropoles by water to their expanding patchwork of colonies beginning in the sixteenth century required improvements in the science of navigation and new understandings and representations of the marine environment. Portuguese, Spanish, and—by the seventeenth century—Dutch, English, and French charts and sailing directions hastened the expanse of commerce and the related growth of naval power. The French Navy established its Dépôt des Cartes et Plans de la Marine in 1720 and the Royal Navy followed, establishing its Hydrographic Office of the Admiralty in 1795. From the earliest portolans to the modern hydrographic charts of the early nineteenth century, navigational science took as its foundational question an understanding of the dimensions of the world and a mission to replace the sea's many mysteries with knowledge. "This is the true description of the world of the cosmographers," read a 1497 Genoese world map based on Mediterranean portolan charts, "accommodated to the marine [chart], from which frivolous tales have been removed." Implicit in such cartographic statements is not just the chart's visuality, but its imposition of narrative—the ability to tell tales—and to remove "frivolous" stories to be replaced by empirical, scientific, and increasingly national ones. The chart filled in *mare incognito*. Where sea monsters once roamed, blank spaces became filled by rhumb lines, depth soundings, and, by the late eighteenth century, the trigonometric angles and baselines that increasingly braced the spaces where waters met dangerous reefs, rocks, and coastlines.[23]

By the early nineteenth century, the world's leading producer of charts and other sources of navigational knowledge was the British Admiralty. Beginning at the prime meridian, the literal and symbolic genesis of longitude at the Royal Observatory in Greenwich, through the development of the marine chronometer by the English inventor John Harrison in the 1730s and 1740s, the British government and its navy understood deeply the interrelation between scientific knowledge and naval, mercantile, national, and imperial power. As Michael Reidy argues in his book on the study of tides in Great Britain during the eighteenth and early nineteenth centuries, "the machinery of empire required as its lubricant a science of the sea, essential for any overseas expansion of trade or successful military campaign." British hydrographers such as James Cook cut their teeth making charts of the St. Lawrence River, for example, which made possible the British capture of Quebec in 1759 that proved decisive to British victory in the Seven Years' War. Later, Great Britain pioneered ocean exploration, as expeditions led by Cook, George Vancouver, James Clark Ross, and John Franklin pushed the boundaries of scientific, navigational, and hydrographic knowledge into new waters and challenging natural environments. In the Royal Navy, observed historian G. S. Ritchie, "naval surveyors combined with the seagoing scientist to form a well-balanced team for the exploration of the oceans." It set the standard for the cooperation of civilian and naval science in the nineteenth century "as a matter of course," a collaboration, for a number of reasons explored throughout the book, that the U.S. Navy struggled to achieve until well into the twentieth. Yet British navigational science stoked commercial but also military and strategic imperatives more than a century and a half before the U.S. Navy embarked on a similar course.[24]

Even by century's end, American hydrographic efforts never surpassed the British Admiralty, whose hydrographic office remained the standard and most comprehensive institution of its kind in the world. Most of the Americans' charts—to the great chagrin of many of its hydrographers—continued to come from Britannia, a fact that reflected the Royal Navy's global dominance during the nineteenth century and the defensive and commercial imperatives of its globe-straddling empire. Surveying and chartmaking was generally an international effort, though, increasingly, by the end of the century, the Americans reserved charts of strategically important waters as secret. Yet for most of the nineteenth century, chartmaking was often cooperative even as it was nationalistic, in part, because the ocean was so vast, deep, and dynamic, and so much work remained to be done, and also because the practice of science itself was perceived as a peaceful, collaborative pursuit even as it stirred so much nationalistic rivalry. Though the U.S. Navy remained wedded to British charts through the turn of the twentieth century, this American cartography nevertheless reflected claims not only to navigation in uncharted or ill-charted waters, but, more broadly, to an emergent sense of identity and nationalism at sea and to an ocean recast first in commercial and then in strategic terms, which reflected the growing interests and changing priorities of the United States overseas. The Americans could not surpass the British Admiralty. Yet neither did the Americans yet aspire to an empire or a navy on quite such a scale.

In the United States, establishment of the depot in 1830 marked an important moment in wresting control and publication of charts and sailing directions from private publishers and marine societies in American maritime communities toward a national, federal system of government-sponsored work located in the U.S. Coast Survey, with jurisdiction over coastal and territorial waters, and the U.S. Navy, with its prerogative to chart foreign waters and the open ocean. Through the War of 1812, the production and dissemination of what historian Matthew McKenzie calls "vocational knowledge" resided in marine societies of port cities such as Boston and Salem, Massachusetts, in some respects, not unlike the earlier leadership of the British East India Company in which Alexander Dalrymple, the Royal Navy's first Hydrographer, began his hydrographic work before the establishment of the Admiralty Hydrographic Office in 1795. In New York, the maritime publisher Edmund March Blunt constructed his American Coast Pilot and published Nathaniel Bowditch's seminal *New American Practical Navigator*. The creation of the civilian Coast Survey in the Treasury Department in 1807, the navy's depot in 1830, and the Army's Corps of Engineers in 1803, reflected a growing, if tentative, role for the federal government in science, which began to replace private efforts with more systematic and professionalized methods and surveying, charting, and publication. Despite the achievements of Blunt, Bowditch, and others, the quality and comprehensiveness of private efforts was uneven. Ocean exploration in the era before 1830 was similarly ad hoc, rooted in private initiatives some of which had unsuccessfully sought federal support. The establishment of the navy's depot, though modest in its beginnings, nevertheless marked an important turning point in the practice of ocean science and exploration in the United States. It formalized scientific knowledge, set standards for the surveying methods used, and sought—though it did not always attain—high levels of precision, uniformity, and comprehensiveness.[25]

Nevertheless, the advent of the depot sparked questions about the place of science within the navy and the federal government more broadly, even as it reflected these general efforts toward government support of those practical sciences such as navigation that seemed increasingly important to the national interest. One of the ironies of naval science is that it had to some degree always occupied a fraught place in the navy's ranks. As important as it proved to the maritime world and, at times, as central as it was in more explicitly military or strategic terms, scientific inquiry long occupied a complicated place within the navy and among an officer corps that in the nineteenth century thirsted for glory in battle perhaps above all else. The Stephen Decaturs and Oliver Hazard Perrys of the navy's rather brief but glorious heritage had not become national heroes swinging sounding leads and taking azimuths. There were exceptions. Thomas Truxtun, for example, who had been master of a Revolutionary War privateer and one of the first captains of the U.S. Navy when it was founded in 1794, participated in some of the first studies of the Gulf Stream when he brought Benjamin Franklin home from London in 1785. Truxtun's biographer notes that the captain, affected by his association with the inquisitive Franklin, "for many years . . . went about plunging thermometers into most of the seas of the world." Matthew C. Perry, perhaps the most influential naval officer between the War of 1812 and the Civil War, immediately began taking soundings of Tokyo Bay when he arrived there in 1852. Nevertheless, as a naval reformer with forward-thinking ideas concerning the potential for new technologies, institutional education, and other professional activities, he was in many ways an exception to the rules common to the American naval officers of his generation. "He who repeats 'by heart' the rules of Bowditch," the naval scientist Matthew Fontaine Maury once groused, "though he does not understand the mathematical principles involved in one of them, obtains a higher number" on his examination for promotion than those officers, like himself, who were "skilled in mathematics" and could "demonstrate every problem in navigation." Maury's quip speaks to the professional challenges facing scientifically minded officers in this era.[26]

Before the establishment of a naval school at Annapolis, Maryland in 1845, education in navigational science and all aspects of the naval officer's profession largely resided aboard ship in the hands of more senior officers. As Maury suggests, a calculating mind for this work was not often nurtured, and certainly it would not put the officer on the fast track for promotion through the ranks. Nevertheless, some officers, such as John A. Dahlgren, the well-known Civil War–era ordinance designer, sought duty in scientific work with the Coast Survey or with the navy because so few officers were inclined to pursue it in times of plodding promotion in a peacetime navy.[27]

The more common practice set down in the *Naval Regulations* that governed the service and the conduct of its officers before a more organized hydrographic program emerged in the late 1830s required passing surveys whenever one's ship spotted an uncharted navigational hazard or came upon one whose existence or location was disputed on existing charts. The report of Master Commandant William B. Finch, commanding the sloop *Vincennes* on a voyage through the Pacific in 1830, is typical:

> In the fulfillment of my orders, I pursued the route most familiar to commerce since the days of the earliest navigators; of course nothing original has been elicited by it in a geographical way. I was not on a voyage of discovery; my instructions were distinct and specific; and the unlooked for extension of an already long cruise forbade delay at any point where I should touch. Yet, professionally, the result is a confirmation, in part, of the remarks and information communicated by Captain Catesby Jones, in so far as our tracks were similar; and the independent ascertainment of the non-existence of Carolina island north of the Society cluster, in the situation assigned to it upon Arrowsmith's chart of 1798, and of two other nameless ones . . . supposed recent discoveries, which are important facts.

Finch's observations and others like it contributed "important facts" to cartographic knowledge, filling in blank spaces on the chart or dismissing alleged dangers that appeared on it but did not seem to exist in reality. Even so, they hardly constituted the kind of sustained program of surveying and charting necessary to serve the nation's commercial and naval needs. As Finch said, he was, emphatically, "not on a voyage of discovery." Hydrography was subservient to other duties, not least, according to Finch, the speedy conclusion of his voyage, which "forbade delay at any point where I should touch." Such reports were undoubtedly important to the operations of the several naval squadrons that protected American commerce around the world, but anything more—such as the sustained trigonometrical surveys required to produce new charts—was well beyond the knowledge of officers such as Fitch, who were not generally inclined to pursue such work.[28]

There were, of course, some promising young officers whom the navy ordered to special hydrographic work in this period, particularly for pressing surveys of the American coast, duties whose practical scientific virtues were readily apparent to a Congress that in this era was slow to embrace an expansive view of the federal government's role in the promotion of scientific activity. Of course, by the 1830s, it seemed clear at least that scientific knowledge would be crucial for westward expansion and for the nation's widening maritime frontier as well.

The federal government's more expansive role in stimulating commerce and economic growth more broadly extended from the National Republicans' American System of subsidized internal improvements that increasingly knit together national markets in a web of canals, turnpikes, and railroads to promoting safe navigation in American and foreign waters. Thomas Jefferson's administration had established the United States Military Academy at West Point and the Army Corps of Engineers in 1802 partly for these reasons. In 1807, he founded the United States Coast Survey to immediately commence a detailed survey of the American East Coast under the direction of the Swiss surveyor Ferdinand Hassler. This work, however, proceeded in fits and starts, as much due to Hassler's prickly sense of entitlement as Congress's hesitance and fickleness in appropriating funds. Nevertheless, during the early periods of the Survey's activity, many naval officers found work in it during times of professional constriction or stagnation in the navy, and under Hassler's tutelage learned the methods and practices necessary to carry out hydrographic work on a sophisticated level.[29]

My own definition, then, of what constitutes a naval scientist in this era is broad, encompassing many different activities and duties from hydrographic surveying according to the trigonometric methods of the day to chartmaking, astronomy, the design and production of new surveying technologies such as deep sea sounders, and more broadly construed theoretical interest in the workings of winds, currents, and all other natural phenomena of oceanic and atmospheric environments. Often, officers came to the Depot or the Naval Observatory with very limited expertise in scientific matters. They left their posts with considerably more. Some sought advancement and promotion in naval science; others came and went, as was the normal case in the Navy's personnel system of that era, merging their scientific duties with the more traditionally accepted role of naval officers as warriors and commanders of the quarterdeck of a man-of-war. Lieutenant Maury, who was perhaps the most important naval scientist prior to the American Civil War and is commonly attributed the title "father of oceanography," was far from the first. Furthermore, in this era—at sea as on land—one need not be academically trained to be a scientist. Most scientifically minded naval officers would have identified themselves primarily as members of a naval—not a scientific or philosophical—profession. The word *scientist* was itself not more than a few decades old.[30]Among the navy's most promising young scientifically minded officers was Charles Wilkes, who had studied under Hassler at the Coast Survey. In 1832, during one of the survey's numerous dormant periods, Wilkes had participated in a navy-led survey of Narragansett Bay. A year later, he took command of the Depot of Charts and Instruments, moving it from its original location on G Street to his own Capitol Hill home, where he installed a small telescope for taking astronomical observations. In 1837, his duty at the depot ended, he led a survey of Georges Bank, a cluster of dangerous shoals off Cape Cod, which had long been a favorite ground of the New England fisheries. The *Boston Mercantile Journal* wrote that the survey "could not have been confided to better hands," as Wilkes was "well known as an officer possessing scientific knowledge. It is to be regretted," the paper added, "that our small vessels are not more frequently employed in service of this description," which "would not only be of vast benefit to our commerce, but would prove of incalculable advantage to our younger officers, who would thus acquire a practical knowledge of the scientific branches of their profession." On the waters of Narragansett Bay and Georges Bank, Wilkes gained experience that in 1838 would secure him command of the nation's most ambitious scientific project to date, the United States Exploring Expedition. Similar in tenor to the *Boston Mercantile Journal* above, this voyage owed much to the cries of the maritime community, which rose in unison calling for better charts of far-off waters now opening to American trade.[31]

Along with the Coast Survey, by 1840 the United States Navy had become one of the nation's most important scientific institutions. While never garnering the unqualified enthusiasm of the service as a whole, the navy nevertheless had the vessels and personnel to carry out a punctuated but sometimes ambitious program of scientific research at sea. By 1844, it would have a proper observatory in Washington that rivaled the observatories in

Cambridge, Massachusetts, and Cincinnati, Ohio, as well as in Greenwich and Paris, the European scientific capitals of the world. In addition to hydrographic pursuits, the observatory, of course, embarked on an agenda of astronomical research championed by Wilkes's successor at the depot, Lieutenant James M. Gillis, who was among the most respected astronomers of his day. Hydrography and astronomy, then, formed the two arms of navigational science in the nineteenth century-Navy. While this study is concerned primarily with the former, it is important to acknowledge that the two were intimately linked in study and practice.[32]

When men like Wilkes looked out to sea and conceived of their work, they perceived the ocean as a wilderness to be tamed by Enlightenment rationalism and by the chart's claims of authority and precision, which rested on a faith in cartographic method and the expectation that the sea's mysteries could be known, its natural forces used to the advantage of American commerce. The history of wilderness in the United States is a complicated one, but cultural and environmental historians, and historians of the West agree that wilderness was a powerful idea in early nineteenth-century America. It served, in part, to justify westward expansion as a place of chaos and disorientation, peopled by savages, inhabited by beasts, and ripe for conquest according to an emergent continental vision in the United States that took as part of its ethos a desire to clear, improve, and thus control the land. Wilderness was also linked, in fundamental ways, to the notion of frontier. In his study of wilderness and the American mind, Roderick Nash wrote that the idea could be defined as, among other things, "the unknown, the disordered, the uncontrolled"—as "a state of mind" in which "a person feels stripped of guidance, lost, and perplexed." Nash was writing primarily about land wilderness, in which American settlers cleared forests of feral beasts and rid it of perceived savages, felling trees, fencing land, and generally clearing the North American forest of the unknown.[33]

By the early nineteenth century, the work of dispelling wilderness became a major intellectual impulse of American expansion on land and also at sea. Nash's definition might as well have been maritime and he, in fact, acknowledged that the term could be applied, in other contexts, both to the sea and to outer space. Historians such as Gary Kroll and Axel Andersson have written about ocean wilderness as it related to twentieth-century marine science. Americans invoked "the western frontier wilderness to understand frontiers like the ocean or outer space," Kroll argues. As Andersson observes, in a postwar Pacific in which the United States sought to exert superpower influence, spectacles such as Thor Heyerdahl's 1947 voyage of the raft *Kon-Tiki* were conceived to "rein in wilderness" and "to control and domesticate the ocean expanses." These ventures, however, were not the first to draw analogies to wilderness and frontier at sea. Rather, such notions were deeply rooted in the ways American hydrographers imagined their work after 1815.[34]

When hydrographers invoked ocean wilderness, they defined it as a chaotic and disorienting nature awaiting the sort of order that their hydrographic charts and texts could impose. Environmental and cultural historians acknowledge that wilderness was not a place. Rather, wilderness was a powerful idea that existed in the minds of these men. Mariners, by contrast, did not often ascribe to it, relying instead on their own ways of making sense of a watery nature briefly outlined above. For hydrographers, however, wilderness served as a counterpoint to the supposed order implicit in cartographic method, process, and representation. In coastal waters labyrinthine in navigational hazards, wilderness was characterized by chaos. Where land met sea, reefs and shoals abounded, breakers crashed in white-capped foam. In the littorals of faraway lands, navigational hazards such as these combined with perceived cultural hazards ashore. The relationship between uncharted coasts and so-called savage peoples was not coincidental. The deep sea, meanwhile, was characterized by a kind of uniformity not present in coastal waters. There, wilderness emerged from the disorientation of feeling placeless, or not knowing where one was without points of fixed reference and only a watery horizon all around. When hydrographers referred to the sea as "boundless" or a "watery waste"—as they often did—they were invoking this aspect of ocean wilderness. Wilderness, they believed, engendered fear and horror, but it also evoked curiosity, the appeal of the strange and mysterious, the lure of the exotic, and sometimes a Romantic reverence for the sublime in nature. It existed at the peripheries of knowledge, in unknown or little-known places, in contrast to the order and knowledge transcribed upon well-charted seas. The uncharted ocean seemed a place utterly without a past and without narratives of empirical meaning. It existed, quite literally, as a blank space on the chart, and thus in the mind. Most importantly, charts constructed vision out of the obscure. They could not materially transform the environment. Surveyors and navigators could not hope to control a littoral awash in breakers in any physical way, but the chart could lend some degree of rationalism to the dynamism of this environment, elucidating safe passages and deep channels through the maze of foam to ride out storms or darkness at anchor in some safe harbor where fresh water, other provisions, or timber for repairs might be found.

In the course of their surveys and chartmaking, hydrographers thus deemed the ocean something akin to the wild North American forests being tamed by American settlement. "And behold!" Passed Midshipman William Reynolds exclaimed in his private journal as the United States Exploring Expedition embarked on its four-year circumnavigation in 1838, "a nation which but a short time ago was a discovery itself and a wilderness, is taking its place among the enlightened of the world, and endeavoring to contribute its mite [sic] in the cause of knowledge and research. For this seems the age in which all men's minds are bent to learn all about the secrets of the world which they inhabit." Here, Reynolds juxtaposed wilderness with enlightenment, knowledge, and research that informed the work of American naval science in this era. He seemed to suggest that the North American wilderness had been civilized, its darkest corners tamed, but the sea remained a blank space—"a waste of waters"—upon which the United States might prescribe its nascent transoceanic visions. The expedition would return to New York City in 1842 with 180 charts, the Fijis among them, and some sixty thousand natural specimens. Wilkes, who commanded the expedition, devoted pages of narrative to Fijian cannibalism, which American readers—the writer Herman Melville among them—consumed with considerable literary appetite. Hydrographers believed that their charts and texts dispelled wilderness, replacing it with the numbered depth soundings and the trigonometric grids that constituted the ideal of an emergent hydrographic science. In 1855, Matthew Fontaine Maury, whose theories the whaler Robert Weir referenced at the beginning of this chapter, likened the mariner with more accurate charts to a "backwoodsman in the wilderness" who "is enabled literally 'to blaze his way' across the ocean; not, indeed, upon trees, as in the wilderness, but upon the wings of the wind." In such oceanic trailblazing, hydrography and the nautical chart were the mariner's axe and spade. Wilderness was the central discursive analogy that linked American expansionist impulses on land and at sea.[35]

Of course, the sea could not be improved—its environment could not be visibly altered—in the ways that Americans transformed the land, and so the particular challenges of this watery wilderness had to be met by a change of mind, that is, by a new way of envisioning the ocean, which, of course, was the purview of charts and a broader thematic cartography that sought to represent different natural phenomena—sometimes, as in Maury's case—with revolutionary new cartographic forms. These, hydrographers believed, would bring order to a largely unknown environment. Controlling the sea, it seemed, was a matter of seeing; it was profoundly visual. "Though the surface is as opaque to [the mariner's] direct vision as it is to that of the landsman," read one account of the activities of the Coast Survey, "the safety of his vessel and all it contains depends upon his ability to see with his mental eye every feature of the subaqueous landscape. This, the Coast Survey enables him to do by providing him with charts." Charts lent just such a mental eye that pierced the surface of the water to "see the waters cleared to such a depth," as Robert Weir put it in his journal of a whaling voyage aboard the *Clara Bell*. Such vision implicit in the chart not only marked trails along highways of maritime commerce set by advantageous winds and currents, but cleared the ocean wilderness of its mysteries, slowly filling in blank cartographic spaces with meaning, narrative, and history, and thus making claims of knowledge and precedence as American commercial and naval vessels pushed the boundaries of oceanic empire. On land, the topographical map was but one tool in which the "improvement" of land dispelled wilderness and recast nature in ways that could be tamed, claimed, owned, conquered, and commodified. At sea, the chart marked the primary means of transformation that was less material than imagined and envisioned.[36]

Wilderness was a powerful literary trope in nineteenth-century America, conjured by American writers from James Fenimore Cooper to Henry David Thoreau and Herman Melville. It was deeply ingrained in the literary aesthetic of the early and mid-nineteenth century. "The ocean," writes Cooper scholar Thomas Philbrick, "embodies most of the qualities that the wilderness does in [Cooper's] novels of the frontier." Cooper's mariner-hero Long Tom Coffin—covered in seaweed, harpoon in hand—is a sort of saltwater Natty Bumppo, a watery trailblazer who for Cooper embodies the individualistic battle with an oceanic nature that, as Philbrick argues, was such an important means of articulating a new American identity for the United States free from British or European models. In his ramblings on Cape Cod, Thoreau mused, "The ocean is a wilderness reaching round the globe, wilder than a Bengal jungle, and fuller of monsters." Edgar Allan Poe wrote of "an unfathomable ocean" and "a wilderness of foam." To Melville, the whale seemed a sort of "moss-bearded Daniel Boone." The whale had "no one near him but Nature herself; and her he takes to wife in the wilderness of waters, and the best of wives she is, though she keeps so many moody secrets." In seeing their own work as dispelling ocean wilderness in the controlled environment set down on the chart, naval scientists drew on the literary conventions of these American writers. It is not a tangential or fleeting connection. Wilderness linked science with literature, chart with text, and the material environment with an imagined one, all of which was integral to the march of American empire at sea. Literary critic Ann Baker, for example, argues in her exploration of measurement in *Moby-Dick*, "Melville accurately saw the embodiment of a relatively new geographical worldview that would enable Europeans and Americans to dominate the globe." As I explain in chapters 2 and 3, even as literary tropes influenced the ways naval scientists conceived of their work, these same writers drew heavily on the charts and texts of naval science and exploration to provide their readers with a more accurate vision of American seafaring and the seemingly exotic people and places at the peripheries of its maritime empire and the liminal spaces of American knowledge and imagination. Science and literature were two ways to make sense of the sea, to reach a broader audience of mariners and landsman readers, and to claim an empire of the imagination.[37]

Like American westward expansion, the concurrent movements of people, commerce, and ideas on the nineteenth-century ocean was an imperial process, and it was one that proceeded fundamentally from the acquisition and representation of environmental knowledge. Oceanic nature—from reefs and rocks to winds and currents and the many mysteries of the deep sea—was a fundamental aspect of seafaring in the Age of Sail. The environment could both destroy life and property and—by fortune, folklore, science, or some mix of the three—hasten voyages to successful, profitable conclusions for American masters, merchants, and shipowners whose vessels ushered an oceanic world forever changed by trade, violence, and cultural encounters. The sea—in both its material and imagined forms— mattered to American mariners. It was existential. And so it is no surprise that an environmental analysis of naval and maritime activities must contend with the ocean, the ways Americans experienced it, understood it, and used it. The rise of an American cartography of empire was at the center of each of these questions. It both served an expanding American maritime world, and, as I argue in chapter 3, it increasingly relied on mariners' close encounter with nature to improve navigation and broader theoretical understandings about natural processes. In all this, hydrographers saw their work as revolutionary, replacing a wild ocean with the boundaries, numbers, and courses of the two-dimensioned chart, thereby expanding the scope of American claims to the ocean.

## 2 Empire of Commerce and Science

....................................................

With all the enterprise of our countrymen, their navy and
commercial marine, still we can say—"Of this huge globe,
how small a part we know."

—Jeremiah N. Reynolds, *Voyage of the United States Frigate
Potomac*, 1835

In March 1839, from a bluff high above the sea in the shadow of Cape Horn,
Passed Midshipman William Reynolds gazed on the Pacific Ocean, ponder-
ing the ocean scape spread before him. "Seated here alone," he recorded in
his journal, "at the very verge of the Western World with a waste of waters
before me that for half the circuit of the Globe rolls on unbroken by a single Isle there was something so imposing in the sublime and solitary nature of the scene, that it seemed to me as if I were like the last man, looking
upon Eternity." Reynolds was a passed midshipman, among the navy's most
junior officers, and a surveyor on the nation's most ambitious voyage of
ocean exploration. As a young naval scientist, he also understood the Pacific as a wilderness that seemed boundless and infinite, drawing on the Romantic tradition in American sea writing that envisioned the ocean in this
era as a frontier wilderness. The men accompanying Reynolds, he observed,
"were stricken and humbled into fear" by "one of nature's wildest scenes."[1]
It might well have been a passage out of Cooper or Poe.

The United States Exploring Expedition sailed at a crossroads of understanding about the natural world. Previous voyages had been ad hoc in their
organization, the initiatives of private individuals who often placed surveying, cartography, or other scientific aims secondary to adventuring and commercial enterprise. The "Ex. Ex.," as it came to be dubbed, was conceived
primarily to stake national claims of scientific and cartographic precedence
to the islands and surrounding waters of the South Sea. When the six vessels
of the expedition departed Hampton Roads, Virginia, in the summer of 1838,
the Pacific Ocean remained unknown in many places, pierced here and there
with half-charted islands limned with reefs. From the Tuamotus to the Fijis,
the indigenous people along these watery peripheries were widely rumored

to be cannibals. The mysteries of the Pacific were the expedition's to dismiss. It charted so-called savage coasts and fixed the biology and cultures of Oceania in scientific texts with titles such as *Mollusca and Shells, Geology,* and the impressive-sounding *The Races of Man.* Many were written by the eminent American scientists of the day. The Ex. Ex. returned to the United States in 1842. One hundred eighty new hydrographic charts and some sixty thousand natural specimens soon followed. One was alive, though barely—a sickly, half-dead Fijian prisoner who had allegedly led an attack on an American trading vessel in 1834. All this bespoke an intense interest in the environments and peoples of the Pacific. To classify nature, to chart the marine environment, and to humble perceived savages was central to establishing an American commercial empire in these waters. The Ex. Ex., in other words, represented an expansive inquiry into the Pacific Ocean, with the ultimate goal of bringing cartographic order to waters newly opened to American trade.[2]

At the expedition's heart were its hydrographic charts, products of a modern cartographic method that fixed the sea in depth soundings and lines of triangulation. From Fiji and the newly discovered Antarctic continent to the Pacific Northwest, these charts constructed coastlines in impressive detail, identifying reefs and other navigational hazards, ship channels, and safe harbors for maritime commerce. Some indigenous ethnonyms remained on them, but the Americans appropriated others, naming islands and harbors after themselves, their messmates, their sweethearts, or their ships—the latter two sometimes seemed almost one and the same. These charts formed an American cartography. They filled previously blank waters with new narratives and meanings that were distinctly national and spoke to American scientific and cartographic precedence. Along with the reefs offshore, the Ex. Ex. also surveyed the indigenous cultures of the Pacific, which these Americans interpreted as complementary specters of an oceanic wilderness that had, over several decades, claimed many vessels and lives. Cannibalism particularly fascinated them—and no surprise—since the numerous stories of shipwreck among uncharted reefs spun in so many sailor yarns usually ended in the mouths of so-called "savages." Environment and indigenous culture merged in the surveys of the Ex. Ex. Its charts and accompanying texts sought to impose control over places *and* people.

In May 1840, after surveying the central Pacific and fifteen hundred miles of the Antarctic continent, the Ex. Ex. turned to Fiji, which was to be a central focus of the voyage. The islands were infamous for their uncharted reefs and the customs of the people. Over the next three months, Reynolds, his fellow officers, and men brought their cartographic visions to bear in a comprehensive trigonometric survey. The expedition's commanding officer, Lieutenant Charles Wilkes, brokered a commercial treaty. Its civilian naturalists collected curious specimens. All observed the habits of the Fijians, who were widely rumored to attack and eat people. These were long, mosquito-infested days, surveying under the tropical sun, sometimes waist deep in rolling surf. Contrary to their hopes and aims, the sea environment did much to undermine the Americans' claims to scientific precision. Then too, the specter of a Fijian attack was never far from their minds. In late July, a group of Fijians killed two officers. The islands, as it turned out, would not be brought so easily into the ordered realm of American empire. Neither did the sea turn out to be a blank cartographic space on which the Americans could impress their visions uncontested. Despite American claims, the Fijians themselves conceived of their environment in quite different ways. When the Americans departed in August, they left three razed Fijian villages and nearly one hundred dead. The military force of this naval expedition was never far behind the peaceful pretenses of science.

The Ex. Ex. struck out into the Pacific at a time of extraordinary change in marine science and exploration. In the United States, voyages of discovery and adventuring had been largely private endeavors—the work of charismatic voyagers, promoters of discovery, and mariners with an eye toward self-aggrandizing narratives in faraway seas—that had received the tacit approval of the federal government, but little more. In the 1830s, the work of ocean science still resided largely in marine societies interested in the promotion of navigational knowledge for the mercantile community and in gentlemen naturalists, many of whom had seized the Enlightenment ideal of the natural philosopher and set out to find, categorize, and write about all manner of natural subjects. In many cases, these men made up in curiosity and exploratory zeal what they lacked in terms of formal education in the natural sciences such as they existed at the time. Narratives chronicling the voyages of John Ledyard, an American who accompanied Cook's third and final voyage to the Pacific, the fur trader Captain Joseph Ingraham, who in the 1790s named islands in the Marquesas after Washington and Franklin, and David Porter's cruise in the *Essex* in 1814 typified the energetic spirit of Humboldtian exploration that inspired many Americans to seek profound experiences in nature and, perhaps more important, to return with fabulous tales of natural, cultural, and geographical curiosities. By 1800, the marine society of the spice-trading captains in Salem, Massachusetts, became a clearinghouse for navigational knowledge and a spectacle in which the curiosities of an exotic Pacific were displayed.[3] Nevertheless, changes were afoot. In the navy, the Coast Survey, and in civilian academic institutions, scientists began to practice in professional settings and according to gradually codified standards of method. Increasingly, scientists came to be supported by and within a federal government seeking to gain some measure of international reputation in exploration and science more broadly, but also informed by a sense that science could serve broader commercial interests, in which the young republic had a significant stake. The voyage of the United States Exploring Expedition marked a watershed moment in these processes.

The course of the Ex. Ex. from its conception to its departure from Hampton Roads in 1838 evinced these changes. Immediately following peace with Britain in 1815, John Cleves Symmes, a war veteran and self-trained natural philosopher, stumped around the country with his theory of concentric spheres, arguing that open polar seas would reveal worlds within the Earth. As far-fetched as the notion now seems, he enjoyed popular appeal and the skeptical attention of the American scientific community as well, reminding us of the limits of geographical and scientific knowledge, particularly at sea, and the relatively fluid lines between different practitioners of science in this era. The polar regions were extreme environments that very few had ever encountered. Those who did were not generally men of science, but sealing and whaling captains driven there primarily in pursuit of profit. Natural philosophers had no monopoly on the construction, production, and dissemination of knowledge about such places. Without the patronage of the federal government, and, in this case, the navy, the production of scientific knowledge at sea happened vicariously through the voyages of commercial vessels and the narratives produced by marine societies on behalf of their captains and merchants.

So potent were Symmes's ideas that they suffused the maritime literature of Edgar Allan Poe in his novel *The Life of Arthur Gordon Pym of Nantucket* (1838) and his short story "MS. Found in a Bottle," the latter published in 1833 just as Symmes's ideas reached their height. In these works, Poe wrote of doomed mariners entering otherworldly environments. "To conceive the horror of my sensation is . . . utterly impossible," observed Poe's protagonist. "Yet a curiosity to penetrate the mysteries of these awful regions," he continued, "will reconcile me to the most hideous aspect of death." In conjuring a mysterious "current" leading directly to the South Pole, Poe suggested that "a supposition so wild" had "every probability in its favor." Men such as Poe and Symmes, in writing and speaking about the limits of

American geographic and scientific knowledge, stoked popular and political interest in exploration with the intention of perhaps validating theories rooted in the intersection between literary fiction and scientific rationalism.[4]

By the 1830s, such curiosity about the mysteries of the sea and the sublimity of extreme environments met the practical concerns of the American maritime community for more accurate navigational knowledge of commercially significant waters and loftier notions of America's place as an Enlightened republic capable of holding its own among the scientific powers of Europe. The newspaper editor-turned-explorer Jeremiah N. Reynolds—who was no relation to Passed Midshipman Reynolds—took Symmes's theory, brushed Symmes himself aside, and called on Congress and the secretary of the navy for a government-sponsored exploring expedition. As historian Aaron Sachs has recently written, Reynolds was an avid American exemplar of the model set by the Prussian naturalist and explorer Alexander von Humboldt. Sachs has characterized the American Humboldtian as gripped by a sort of wanderlust to experience nature, to place himself—the scientist—in it, and to see the natural world as an interconnected whole. Reynolds had organized an abortive private venture to search for an Antarctic continent—Terra Australis—in 1829 with the sealing captains Nathaniel Palmer and Edmund Fanning. A decade later, based on that experience, he wrote *Mocha-Dick; or, The White Whale of the Pacific*, which Herman Melville read with interest. He had also published a well-received book on the Pacific voyage of USS *Potomac*, whose commander, Captain John Downes, served as David Porter's second-in-command aboard the *Essex* during the War of 1812 and who, during the *Potomac*'s voyage, razed Kuala Battoo on the island of Sumatra, one of the most significant examples of American gunboat diplomacy in the Pacific before the Civil War. Reynolds was deeply immersed in the American commercial and naval worlds of the day. An American expedition to the South Pacific, he thought, might encounter open polar seas or perhaps a new continent. Meanwhile, the Americans would survey vast stretches of uncharted Pacific waters for whalers and merchant mariners who, according to Reynolds, "set forth on untried voyages, and *cut* with daring keels unknown seas with the same hardihood that impelled them to pierce unexplored forests, and tame the howling wilderness" [italics added]. Hydrography proved a compelling enough political and commercial argument where invocations of hollow earths did not. Here, Reynolds too invoked the idea of oceanic wilderness so prevalent in literary circles of the time to make a case for an expedition that would assist mariners in taming a wilderness of the sea. "To cut" was Reynolds's operative verb here, whether it be paths through the sea or the forest. For advocates like Reynolds, it was the chart that could best aid America's mariner-pioneers to press their keels into unknown seas.[5]

Reynolds joined other voices in the maritime community—the East India Marine Society of Salem prominent among them—calling for federally sponsored voyages of exploration and survey that would provide masters and mates of commercial vessels with more accurate charts. The appeals arose like a chorus from Salem, New Bedford, and Nantucket, the maritime capitals of American seaborne commerce, whose mariners were the frontiersman of American sea literature. The sealing captain-turned-explorer Edmund Fanning testified before Congress in 1833 concerning a Pacific trade worth "hundreds of thousands, aye! Millions of dollars." Reynolds wrote to Secretary of the Navy Mahlon Dickerson that mariners needed "national protection," which he defined first as "protection from the dangerous reef," and, he added, "also from savages." In Reynolds's mind, the two were constitutive elements of oceanic wilderness. He wrapped all this in the glory of scientific discovery. The expedition would represent a distinctly American contribution to knowledge vis-à-vis Europe. The Pacific, Reynolds told Congress in 1836, was "truly our field of fame." He regarded exploration as an imperial act, one calculated to stake national claims through heroic feats of discovery amid savage environments from the reef-strewn central Pacific to the ice-bound shores of what he and others supposed must be a Terra Australis. It was Reynolds, then, perhaps more than anyone, who nationalized American ocean science and enlisted it in the cause of the nation. He ushered it from the hands of the amateur explorer—of whom he himself was one—to a project that held tremendous significance for a nation beginning to articulate its independent identity and its commercial and scientific ambitions. The sea, Reynolds thought, would be the space and the environment on which Americans should make their claims.[6]

Ultimately, the expedition would be an awkward marriage of mid-nineteenth-century science—civilian and naval, natural and hydrographic—that reflected the ongoing transformation of the discipline, the common but tenuous interests of various, diverse groups, and the diverging identities of multiple communities as the sciences became gradually more specialized and professionalized. The navy and the nation had mounted nothing like it before. It surpassed in expense, size, and complexity of the Lewis and Clark Expedition through the West, 1804–1806, though Reynolds was keen to point out parallels between these two efforts. "One went into a wilderness almost entirely unknown to our people," he observed, "and the other was to go into seas that were partially known to them . . . but whose rocks and reefs were not known."[7] Through the 1830s, the expedition's fate experienced fits and starts, the result of congressional fickleness and, not least, the navy's own uncertainties at the prospects of such a project.

Among other things, before ultimately settling on Lieutenant Charles Wilkes, the navy could not find a willing commanding officer either to abide the long delays and uncertainties about the expedition's fate, or who evinced enough specialty in and enthusiasm for science broadly construed within a service that did not place much professional value in such pursuits. "An enterprise of science should be in the hands of science, to which I lay no claim, and for which I have sought not more than became necessary for the common management of vessels under my command," wrote Captain Lawrence Kearny in the *New York American* in early 1838 as his name was being floated for the expedition's command. To Kearny, science seemed "no easy task to perform. When commercial interests are involved, I can see my way clear . . . but when the wings of *science* are spread, they soar too high for me to reach a feather." Kearny's admission speaks to the limits of his expertise and that of many of his peers in the naval officer corps. As officers, they understood keenly the young navy's commitment to furthering American commercial interests at sea and over foreign, sometimes hostile, shores, but any expedition that also aimed to a larger scope of scientific inquiry into the natural world seemed too alien an undertaking.[8]

Indeed, for the navy during this era, science was to be strictly defined in navigational and cartographic terms, one corner of an expanding marine science that naval officers such as Wilkes deemed important to national interests and one that naval officers like himself would be qualified to execute. Whatever the expedition might become—to men like Reynolds and to political allies such as former president John Quincy Adams, who had his eyes on national universities and national observatories that would make the United States a scientific republic—the Ex. Ex. would be defined, first and foremost, as a naval expedition conceived to survey the waters of the South Sea and to produce charts for the navy and the American commercial fleet. As Reynolds proposed a "corps of naturalists" drawn from civilian science who would, in his words, "study the organic world alive," sending gentleman "scientifics" such as the geologist James Dwight Dana, the philologist Horatio Hale, and the naturalist Charles Pickering to sea, he wrangled with successive navy secretaries, and ultimately with Wilkes, the expedition's new commander, who was committed to running a naval expedition for naval purposes. Wilkes later admitted that he would not countenance an effort in which his officers were left to be "hewers of wood and drawers of water only." No, hydrography would be its primary aim, the "essential objects of the Expedition." These "duties that were enjoined upon me," he wrote, "were entirely unknown to [the scientists]." Thus, from the outset, the expedition was both an unprecedented amalgam of the American scientific world and a fragile and discordant mash-up of competing interests and views. With "High Admiral [Jeremiah] Reynolds and the various tribes of 'Logos,' " the expedition "will resemble . . . a caravan of fiddlers, idlers, etc.," one officer derisively told Lieutenant George Foster Emmons, referring to the seven naturalists who would accompany the expedition. "Make your own notes and write your own books," Emmons's friend counseled, "otherwise these scientific gentlemen will be sure to figure in the histories . . . as the principal personages."[9]

When the Ex. Ex. finally departed from Hampton Roads in August 1838, bound for the Pacific, it carried the weight of the nation's scientific reputation and the prospects of its growing commercial empire with it. These were "gallant men, who now sally forth on the errand of the country," reported one newspaper as the Ex. Ex. sailed over the horizon. "And wherever floats that standard sheet—the star spangled banner," it declared, "over brave men engaged in the public service, thither will the sympathies and affections of the people tend, and will encourage, and uphold, and reward." In his orders to Wilkes, Secretary of the Navy James K. Paulding emphasized that the voyage would be "for the purpose of exploring and surveying" to "determine the existence of doubtful islands and shoals" and "to accurately fix" them on the expedition's charts. The voyage "necessarily attracts the attention of the civilized world," the secretary reminded Wilkes, "the honour and interests of [the United States] are equally involved in its results." Paulding conceived of the voyage as extending an American "empire of commerce and science," by which he meant expanding American commercial enterprise across the ocean and claiming the mantle of scientific and geographical discovery. Great nations contributed to knowledge about the natural world. In the Ex. Ex., the United States now sought to more fully join Europe, the undisputed center of Enlightenment science. The voyage would be not merely national, but nationalistic.[10]

The expedition's surveys and its vast collections of natural specimens set down after the voyage in a hydrographical atlas, a five-book *Official Narrative*, and seventeen volumes of scientific texts would be more than aids to navigation and scientific treatises. Paulding, echoing Reynolds, had infused them with potent meaning and power tied to the honor and reputation of the United States and its commercial interests. This empire of commerce and science was Manifest Destiny gone to sea. It was not secured by war or treaty—though the Americans did resort to violence and diplomacy in the pursuit of their antebellum imperial goals—rather, this would be an empire defined by the peaceful pretenses of knowledge set down in the charts and texts published in the expedition's wake, the work of American surveyors armed with potent ideas about the transformative power of cartography to "fix" or "fix-in," as the Americans often put it, a chaotic, dynamic, and largely unknown ocean wilderness. In Paulding's vision, the expedition's work would be nothing short of American trailblazing. Its charts, he told Wilkes, would "enable future navigators to pass over the track traversed by your vessels, without fear and without danger."[11]

Historians have long discerned in the expedition's voyage, its surveying, and its other scientific endeavors an important moment in antebellum American empire. From William Goetzmann to D. Graham Burnett, they have emphasized the significance of ocean exploration and the methods and practices of hydrographic surveying as imperial acts. Ann Fabian and Antony Adler, meanwhile, have examined the Americans' interest in cannibal bodies as a means of intellectual and cultural conquest. The Ex. Ex. "ushered the nation into an Enlightenment tradition of imperial voyaging," historian of science Michael Robinson has recently written. Yet historians have not fully examined the ways in which science and military power converged in the natural environment in the service of imperial visions. That cartography could seem to impose order and rationality on a wild marine environment was central to Americans' visions of control, which proceeded in this era less by diplomacy and military force than through invocations of cartography and scientific inquiry.[12]

As contentious as the expedition's scientific aims were, its vision of a Pacific recast according to American aspirations for commercial empire rested on commonly held assumptions about the growing power of science— hydrographic or natural—to bring order to a little known, chaotic, and sometimes savage wilderness through precise method and practice. Whether it was the capture and categorization of South Sea flora and fauna, or the precise angles of the trigonometric survey, the expedition's transformative power lay in the ideal of precision. "Would it not be honourable to our national character," Jeremiah Reynolds had asked, "if the observations and collections made through the agency of our own naval and scientific intelligence should be distinguished above all others by their accuracy and completeness?" As the expedition rounded Cape Horn and entered the Pacific Ocean in March 1839, Wilkes drew up orders for his officer-surveyors, directing them in the practice of the hydrographic survey, which he had learned under his mentor Hassler at the Coast Survey. "The reliance to be placed upon hydrographical labors," Wilkes reminded his men, "depends upon the accuracy of the modes employed in obtaining the results." Accuracy alone would garner America a seat at the table of scientific nations, and it alone would save mariners from a grisly fate hard aground on coral reefs, overcome by supposed South Sea cannibals. It was the expedition's fidelity to the cartographic ideal of precision and comprehensiveness that, among other things, marked its departure from earlier American exploratory efforts at sea and promised, according to the worldview of its scientist-surveyors, an ocean recast for American voyaging.[13]

The surveying process was itself an act of control over nature, over the surveyors' own bodies, and over others—both American and indigenous. Wilkes's directions had his officers command small whaleboats or launches of six to ten sailors, powered by oars or sails, and stocked to the gunwales with compasses, sextants, theodolites, and signal flags. The trigonometric survey evinced "precision of measurement and skill in instrumentation," which historian of science Robert V. Bruce has deemed the "two related American scientific tendencies." Working in pairs, with a larger ship of the expedition or alone, these boats wove their way among the shallows offshore. The officers measured horizontal angles by sextant, sounded for depths using a measured lead line, and calculated distances by the speed of sound determined by the shot of a cannon or blunderbuss between fixed points. Returning to the ship after a day's or sometimes several weeks' work, the officers connected the various subsurveys, checking by trigonometric calculation the exact distances between points with known angles. Wilkes anchored the whole by a central position of latitude and longitude usually taken from a makeshift observatory. Uncharted or so-called imperfectly charted islands thus emerged bracketed in triangles that spanned the dangerous littoral from ship to shore, linking together land and sea in the dangerous and ever-changing spaces where these environments converged. Of course, it was also at the convergence of land and sea environments where ships came to trade, seek refuge, gather a commodified nature of sandalwood or sea slugs, and, sometimes, where they met their fate on uncharted reefs or at the hands of hostile indigenous people. Accurate charts predicated in the trigonometric survey gave navigators some control over all these outcomes.[14]

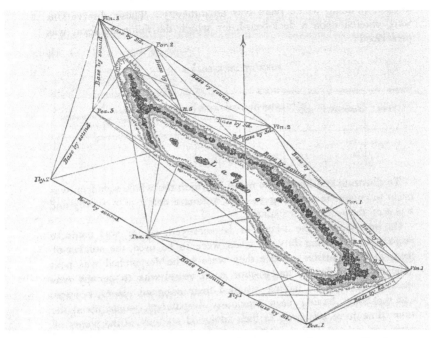

Diagram of the United States Exploring Expedition survey showing Wilkes's trigonometric method with baselines and lines of triangulation, from *Narrative of the United States Exploring Expedition*, vol. 1. Courtesy of the Beinecke Rare Book and Manuscript Library, Yale University.

The officers' journals evince the centrality of surveying work, which linked their daily experience to the Americans' broader mission, but also the ways the practice of hydrographic surveying could reimagine ocean environments. These men recorded their work in methodical, repetitious, and exhaustive detail, not so much to impart knowledge, or to describe their daily experience only, but also, one gets the sense, from a sort of self-perpetuating belief that attention to method and discipline was a productive end in itself, that is, not just in their surveying work but also in the routine of recording these activities in their journals. It all became part and parcel of the habits needed to bring order to the Pacific's ocean environments. An excerpt from Lieutenant George Foster Emmons's journal in which he records the expedition's survey of the Tuamotus in August 1839 is typical: "Took our station for surveying the islands, this ship and the schooners to windward and the *Vincennes* and brig to leeward. Measured as usual by measuring base by sound as soon as we had obtained our positions, then taking azimuths etc." "The work was excruciating and unceasing," recalled William Reynolds. "Surveying keeps the mind on the utmost stretch of attention and a moment given to relaxation might spoil or injure the whole. There are a thousand things to be noticed, both near at hand and at a distance, and the employment of all the faculties is necessary to an almost painful degree." The whole process of surveying was one of correct bearing, D. Graham Burnett argues, to which naval discipline and the hierarchy of command lent itself particularly well. As Reynolds suggested, to impose order on environments, the Americans had first to impose order and discipline on themselves.[15]

The survey's attention to detail and precision was the hallmark of Wilkes's cartographic method, the point on which the survey's transformative power turned. For these men, the trigonometric survey imparted what cartographic historian Matthew Edney has called "epistemological certitude." It transcended the quantification of nature to the supposed dominion of it, convincing cartographers that "the world can be mapped exactly, the world can be *known*." For Edney and many others, maps are powerful ideological instruments of empire not only represented by the finished charts themselves but implicit in the process and practice of cartographic method. The same was true of sea charts, which in coastal waters relied on the same general surveying methods. D. Graham Burnett has argued that the expedition's charts represented powerful tools of "imperial transformation." Their construction required a particularly demanding kind of mental acuity—a deep intellectual engagement with the natural world—that facilitated this imagined turn from disorder to order. In observing and measuring the littoral environment, this work was also intensely visual. It was through the trigonometric survey—its spyglasses, theodolites, and sextants—that the Americans cast their imperial gaze, transforming wilderness into commercial empire by measuring and then transcribing their field of vision onto the chart's field of representation so that future mariners might see what existed beneath the water's surface.[16]

Even as they sought to fix reefs and islands on their charts, their eyes were also fixed on the people who inhabited these islands. Culture and environment were deeply intertwined in the ocean wilderness. "Savage coasts," as at least one mariner-writer deemed them, were neither strictly nature, nor culture, but a commingling of the two. The Americans, then, could not think about coral reefs without eyeing warily the people ashore. By the same token, the habits of the indigenous people of Oceania took on specters more fearsome by the navigational hazards offshore. On a more prosaic, but no less important level, shipwreck on uncharted reefs cast mariners into the "wild" hands—and, supposedly, the mouths—of indigenous peoples.[17]

As the islands of the Tuamotu group appeared on the draft charts in the late summer of 1839, the officers felt compelled to fill their journals with page after page of anthropological description. As Emmons put it, again, with emphasis on the visuality of exploration, the men were filled with "a laudable curiousity to see places that they never expect to see again." They became interested in humans as much—perhaps more—than reefs and harbors. And the interest was mutual. "They watched our movements with great attention, sometimes squatting down upon the sand and gazing at the ships for an hour or more without moving or changing their position," recorded ship's surgeon Silas Holmes of the Tuamotuans, whom the Americans simply referred to—through their own frontier lens—as "Indians," for it was the Indians who inhabited the North American wilderness; it was the Indian who seemed to be falling before the U.S. Army and the American settler. Emmons, for example, noted that the Tuamotuans "appeared to be going through drill-marching in Indian file." He described one man as "occasionally jumping at *right angles* with our *line of sight*" [italics added]. In so many angles and lines of sight, he could not help, perhaps, framing indigenous cultures in the same cartographic terms as the surrounding waters. Indeed, time and again, the Americans turned from cultural observations to cartographic ones, as if the people of the Pacific were one more unknown quantity to survey. "The countenances of all appeared highly animated and they were evidently much awed," Emmons observed. He then returned immediately to the business at hand. "Found the thermometer at 550 fathoms to stand at 45 degrees—compared chronometers—at meridian the land bore from 55° E to S 80°." Environment and culture, then, were inseparable. If cartography was in large measure about ridding the sea of its wildness, then the Americans were also impelled to turn their cartographic gaze on the Tuamotuans and others.[18]

A completed chart of the Tuamotus, which ultimately appeared in Wilkes's five-volume *Official Narrative of the United States Exploring Expedition* in 1844 for mariners as well as a broader American readership, shows the ways in which environment and culture were inextricably bound together. On the chart, the Tuamotus are bisected by two dotted lines. One marks the farthest reach of Euro-American missionizing. The other identifies those islands in the group still thought to be mired in cannibalism, as if Christianity and cannibalism were mutually exclusive practices on these islands. The chart functions here as a malleable tool of American commercial empire,

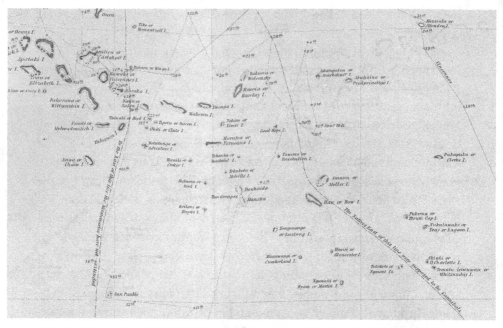

Detail of *Map of Low Archipelago or Paumotu Group* charting the influence of Christian missionizing and cannibalism, from *Narrative of the United States Exploring Expedition*, vol. 1. Courtesy of the Beinecke Rare Book and Manuscript Library, Yale University.

capable not only of saving a ship from wreck with the total loss of property and life, but also saving the shipwrecked themselves from a grisly fate at the mouths of supposed cannibals. That cartography held the power to delineate culture—an amorphous thing if ever there was one—speaks to the assumptions that pervaded American naval science and the broader set of beliefs that underscored American commercial and cultural imperialism in this era. Charts were primarily tools of navigation. Yet the Americans invested them with powers not only to represent a marine environment in constant flux but to document the process of cultural change as the Pacific entered the Euro-American cartographic and literary imagination. On charts like this, the Americans' cultural survey of the Tuamotuans took cartographic form.[19]

For the expedition's civilian naturalists—some of whom had been assigned to the expedition for the express purpose of studying Pacific cultures—the voyage had become a frustration, their work mired in the shadow of Wilkes's hydrographic priorities. Left largely to view the Tuamotus from the expedition's ships, the civilians chafed at Wilkes's authority and the challenges of doing science as civilians aboard naval vessels at sea. Clearly, there was tension between some of the expedition's officers and the scientists. At the outset of the voyage, Wilkes ordered his men to treat the civilians as "one of us," pledging that "my conduct towards them will be the same as towards the officers with whom they are associated." Yet Emmons could not help snickering at their strange, unmilitary habits. "Nothing else occurred worthy of note," he recorded on one particularly uneventful day, "unless it be the appearance of one of the Scientific Corps upon the quarter deck with an umbrella spread." The work of the scientific corps did interest men like Surgeon Silas Holmes, whom the botanist William Rich harangued to collect specimens while he remained incapacitated by sea sickness. Passed Midshipman Reynolds "day's amusement" was often spent observing the scientifics "scoop nets into the ocean to catch the Animalculae or Verterbrae with which the water is alive—It is wonderful to observe," he concluded. But it was also obvious that these men cast a strange figure on the deck of a man-of-war. As Rich's misfortunes illustrate, the scientists often found the sea environment more challenging to scientific inquiry and more foreign to gentlemen of science than the more familiar halls of Yale College or the meetings of the American Academy of Sciences. Indeed, one of the most fundamental challenges for nineteenth-century marine scientists was simply going to sea at all and remaining there long enough to gather valuable knowledge.[20]

Nevertheless, despite the difficulties of conducting science at sea, Wilkes's heavy-handed treatment of these "scientifics," and his hydrographic focus, the expedition represented a watershed moment for American science more broadly. In this odd marriage of naval and natural science, the two met mutual needs. The scientists lent a sort of credibility based largely on the model set by the British Admiralty and the example of Cook and others who followed him going to sea with at least one naturalist aboard. Meanwhile, federal patronage of science in the form of navy-run expeditions promised to take civilians from academic settings to oceanic ones where they might actually encounter, and thus experience, natural environments and their flora, fauna, and people in ways commensurate with the era's focus on Humboldtian science. The well-known geologist James Dwight Dana, for example, derived from his circumnavigation of the Pacific an emerging theory about the ways in which coral reef formation was tied to the drift of continental plates along a geologically constructed Pacific Rim. According to historian David Igler, Dana's work and his book *Geology*, one of seventeen scientific volumes edited by Wilkes to be published after the expedition's return, at first advanced a unique understanding of what would become plate tectonics, in which Dana framed a Pacific not foremost in expansionist terms as the westward march of American civilization, but rather as the product of geological forces that connected a Pacific World that happened to include places such as Oregon and California. Nevertheless, Igler contends, by the 1850s, Dana's constructions of America's Far West had given way to a brand of science "that celebrated this path of imperial consolidation," seeing America's West Coast linked to a larger vision that looked to Hawaii and Japan. Dana's colleague, the naturalist Titian Peale, meanwhile, engraved in silver on the stock of his musket, the location and date on which he had fired at indigenous people. Altogether, the Ex. Ex. collected some sixty thousand natural and anthropological specimens, which went on display at the National Gallery in Washington, ultimately forming the basis for the collection of the Smithsonian Institution. Thus, while the scientific corps' aims and practices remained strange and bemusing to the expedition's naval officers, both hydrography and natural science ultimately proved to be mutually reinforcing scientific arms of American expansion during this era.[21]

When Wilkes turned his ships southward toward the pole in December 1839, it was to seize the reins of a nationalistic rivalry in science, exploration, and discovery, to claim American imminence on this oceanic stage of empire, and to press American claims to Antarctic discovery. The expedition's two Antarctic forays—the first had been in early 1839—were as much about science as they were about discovery and glory for the United States. In March 1839, Wilkes and his men had attempted to surpass Cook's "Ne Plus Ultra," the highest southern latitude reached by the British explorer in his quest to reach the South Pole before being thwarted by ice on January 30, 1774. The name itself evoked the unknown and barriers meant to be broken. Now the Americans hoped to best Cook and, by extension, British scientific claims to the South Sea, first claiming legitimacy for themselves by following in the great explorer's wake, and then seizing the glory of discovery by surpassing him, a technique of exploration that historian D. Graham Burnett has called *metalepsis*.

To the Americans, Cook was the pole star by which they navigated their own visions of exploration and by which they continually compared themselves. Many of the officers had read extensively from Cook's voyages and in the annals of exploration and discovery more broadly to orient their own historical place on the sea. Since departing Hampton Roads, Passed Midshipman Reynolds had immersed himself in "the histories of English discovery and surveying voyages." It was "the immortal Cook" that most inspired him, "penetrating so far south as he did, among the Icy seas—running his vessels where Christians had never been before—into parts entirely unknown, where dangers and difficulties unseen until they were felt, met him at every step." Reynolds and the Americans sought to be immortals, too, which they hoped to achieve by surpassing Cook and by testing themselves against an extreme environment where navigational prowess in uncharted seas made for stirring narratives and, by extension, national claims.[22]

The pursuit of Cook's Ne Plus Ultra reads as one long nationalistic rivalry set amid the icy seas of the Southern Pole. "Every pulse now beat with emulation," surgeon James Palmer aboard the schooner *Flying Fish* remembered, and "hopes began to brighten at the thought that they had passed the French and Russian limits and were on the heels of Cook." But the environment interceded. "Nature had let loose all her furies," Palmer reported to Wilkes. The Americans were compelled to return north short of Cook's mark. "Our hopes were blasted in the bud," Palmer reported. Cook's Ne Plus Ultra—seemingly little more than some icy spot on the sea—took on quite powerful meanings for the Americans who sought to assert their growing independence from Europe and to claim the mantle of discovery for themselves. Euro-American exploration set down on charts and in texts such as Palmer's Antarctic prose poem *Thulia* and ultimately in Wilkes's *Official Narrative*, imparted a set of stories to these waters with which to assert or contest national claims. From Jeremiah N. Reynolds's original vision, the Ex. Ex. sought polar voyaging amid an extreme environment as a way to interject America's growing influence along the geographical and imagined peripheries of an oceanic commercial empire, ripe, as Wilkes later reported, in seals and whales. In early 1839, the Americans returned northward unsuccessful. They had not bested Cook and the British. Rather, they had been bested themselves by an Antarctic environment that tested the limits of human endurance.[23]

When the Ex. Ex. returned in the austral summer of 1839–40, the Americans discovered land of continental proportions, a claim contested by contemporary French and British expeditions also searching for a supposed continent, but one that Wilkes pressed primarily through cartographic evidence. Aboard USS *Peacock*, William Reynolds had been among the first to spot snow-covered mountains looming above the icy barrier in January 1840. It was a continent, confirming reports going back to the early nineteenth century that some land existed in high southern latitudes. But how much? The Americans determined its proportions after surveying fifteen hundred miles of Antarctic coastline. It was a claim, as the Americans learned on their victorious return to Australia, that contemporary European expeditions under Dumont d'Urville and James Clark Ross soon contested. Ross, who commanded a British expedition and was a famed polar explorer in his own right, refused to give precedence to the Americans, having traversed the Americans path set down on their chart, and finding no land where Wilkes had placed it.

Wilkes, however, invoked the chart as a testament to the Americans' precedence and accuracy of their claims, but also to narratives of struggle and perseverance against an icy polar environment, and to daring feats of navigation amid moving bergs and ice flows. The Americans believed authority derived from the chart whose accuracy was beyond all doubt, a claim ultimately upheld by later explorers. "Examine all the maps and charts published up to that time," Wilkes challenged, "and upon them will any traces of such land be found?" On the Americans' chart, by contrast, a vast continent emerged. There too, the paths of the American vessels appeared weaving in and out of icy seas. The chart marked Disappointment Bay, where initial visions of rocky land had proved elusory, and Peacock Bay, named after Reynolds's much-loved ship, which had nearly gone to pieces pressing southward. Farther inland, Reynolds and Eld peaks marked the moment and place of first discovery, and the discoverers themselves. This toponymy not only proved American claims but imbued the environment with narrative meaning that testified to the Americans' experience there.[24]

Charts, then, proved the most effective means of staking and supporting claims to discoveries in a Pacific being transformed by Euro-American scientific, economic, political, and cultural influence. Upon the expedition's return to Sydney, and facing suspicion and contempt from Ross, Wilkes promptly left a copy of his chart for Ross to consult as the latter headed southward the following year. For Wilkes, the chart had established "truth in relation to a claim that is indisputable. How far Captain Ross was guided in his search by our previous discoveries," Wilkes continued, "will best appear by reference to the chart. There can scarcely be a doubt of the existence of the Antarctic continent," he wrote to Secretary Paulding, though he thought it "unbecoming" to "speak of our arduous services." With his penchant for immodest understatement, he would let "the report and accompanying chart of our cruise . . . speak for us." Where in early 1839, the Americans had followed Cook's wake, the following year it would be Ross, Wilkes intended, who would follow with the American expedition's chart in hand.[25]

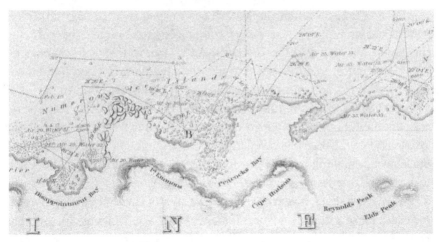

Detail of *Chart of the Antarctic Continent, 1840,* showing the tracks of the expedition's vessels. It includes Reynold's and Eld's peaks, where the Americans first sighted land on January 16, 1840, and Disappointment Bay, where the Americans had been turned back by ice. Courtesy of the Library of Congress, Geography and Map Division.

In May 1840, the Ex. Ex. arrived in the Fiji Islands to conduct a three-month survey that—after the discovery of Antarctica—was to be the expedition's crowning achievement, forging an American commercial empire out of perhaps the most infamous ocean environment in the Pacific. A survey of the Fijis had long animated calls for a national exploring expedition. In 1837, the expedition's staunch promotor, Jeremiah N. Reynolds, had published in the New York press an open letter to the secretary of the navy, highlighting "no inconsiderable trade in native productions. The amount of property lost there," he reasoned, "would pay a goodly portion of the expenses of the expedition." Spread across some 95,000 square miles of the southwest Pacific, the Fijis consisted of approximately 150 islands and 80 reefs by Wilkes's count. Many of them were uncharted or ill-charted, which had discouraged explorers from the time of Tasman and Cook. But American mariners and merchants would not be turned away. They were first lured by sandalwood, which merchants sold in China for use in ornamental carvings and as incense. By midcentury, however, Fiji's sandalwood groves were nearing exhaustion. Later, merchant captains, most of whom hailed from Salem, Massachusetts, established a lucrative trade in bêche-de-mer, the reef-dwelling sea cucumber *Holothuria,* much desired in the China market as an aphrodisiac. American whalers also called there to reprovision, and so when the Americans arrived in 1840, a small Euro-American community of castaways, traders, and missionaries lived among the Fijians. "The Natives look upon the Americans as their natural customers," William Reynolds observed. Apparently by their own account, Fijians knew well the "character of Napoleon and Washington," and the print of "General Jackson was highly valued." One Fijian chief, Silas Holmes recalled, "knows our flag and says that vessels often visit this part of the group." When the Ex. Ex. arrived, then, Fiji was already linked to the United States in a deepening commercial and cultural relationship forged by Americans' growing presence among the islands.[26]

The reefs offshore, however, presented a significant impediment to trade. Fiji was a maze of them, and most remained uncharted. Lieutenant Emmons conveyed the environmental dangers facing navigators. "The strength of the wind and the short combing sea, the latter being produced principally by a strong current which is most everywhere apparent around these islands and added to the numberless shoals that encompass them makes the navigation extremely dangerous." The expedition's working chart, which Wilkes had compiled from various word-of-mouth and cartographic sources, was "a frightful display of rocks and reefs garnished here and there with notices such as 'brig Eliza lost'; 'Am[erican] brig lost;' etc. etc. etc.," William Reynolds noted grimly. Even as the Americans began their survey, the whaleship *Shylock* struck a reef, prompting Reynolds to write home that "the number of Americans and other whites that have been destroyed by the Fegees [*sic*] is so great, that if the Islands were to be depopulated entirely the retribution would not be enough to repay the loss of life and property." In the Americans' mind, the expedition's survey and the subsequent charts would bring order to these waters, saving lives and property, and more firmly bring the Fiji Islands into the realm of America's commercial empire.[27]

Yet uncharted reefs were only one part of a fearsome marine environment amplified by the perceived savagery of the people ashore. The littoral convergence of land and sea—the very natural spaces the Americans hoped to survey and rationalize on their charts—had been the setting of so many shipwrecks and supposed cannibal repasts. In promoting an expedition, Jeremiah Reynolds identified the Fijis as islands "not one of which can be found . . . on any map or chart hitherto published. More than one hundred . . . American seamen," he claimed, "have been shipwrecked, and a large majority of them sacrificed to the murderous cruelty of the natives at the Fiji Islands alone." Ship owners could not obtain insurance on vessels bound for Fiji, not only because of the navigational hazards, Wilkes reported, but because of "the native character. A thorough knowledge" of it "is essential to success," he concluded. The loss of the Salem bêche-de-mer brig *Charles Doggett* in 1834 in which Fijians killed and apparently ate eight crewmembers was still fresh in the Americans' mind as they entered the islands. The Fijis had a reputation; they were a public focus of the expedition even before it left the United States. In fact, Congress and the Van Buren administration had been spurred to act on a surveying expedition, in part, by so many memorials from the maritime communities of Boston, Nantucket, New Bedford, New London, and Salem. Americans could not talk about Fijian reefs without talking about cannibals, and they could not speak of the Fijis without reference to both.[28]

Whatever questions the Americans might have had about the cultural habits of the Fijians, their minds seem to have long been influenced by voyage narratives, sailor yarns, and, not least, by the fearsome marine environment that ringed the islands in jagged coral. The scholarship on Fijian cannibalism is deep and contested. The anthropologists William Arens and Gananath Obeyesekere have argued that the Fijians' reputation for cannibal feasts informed the Americans' minds from the start perhaps more than any actual observance of them. To this, I would add that the specter of the marine environment around which the Americans' hydrographic survey revolved was particularly influential. As Paul Lyons has observed, the Americans regarded Fiji's reefs "as the natural equivalent of human dangers." The surgeon Silas Holmes, for example, found his observations "in perfect harmony" with what he had already heard, while Lieutenant Emmons noted that he did "not like these people," deeming them "ugly, disgustingly dirty, and cannibals besides." For the newly promoted Lieutenant William Reynolds, the Fijian wilderness constituted both navigational and cultural dangers. "The people are generally believed to be ferocious cannibals," he observed, "and the numerous reefs and shoals and labyrinths of rocky passages among the cluster are so many snares for the seaman's destruction." In his mind, the "treacherous nature of the people" would be "added to the perils of navigation." At the end of May 1840, he sighed, "well we are among the Feejees, and have not been killed, nor eaten, nor wrecked yet." Here, Reynolds drew explicit lines between Fijian reefs and Fijian cannibals. The fearsome marine environment seemed to convince the Americans that the Fijians actually ate people. Conversely, rumors of cannibalism ashore only amplified navigational dangers across the beach.[29]

It is no coincidence, then, that surveyors like Reynolds and others—indeed, virtually every officer that has left a record of his participation in the Fiji survey—harbored a deep interest in the question of cannibalism, suggesting that their pursuit of such inquiries was not just tangential curiosity or some morbid fascination, but rather part of a deeper study of the oceanic world in which hydrography and anthropophagy were intertwined. "The natives began to appear on the beach," Reynolds wrote as the Ex. Ex. arrived in the islands, "and we took our first look at the Feejees—the *survey* was unsatisfactory—for they were ill-looking beyond conception" [italics added]. Here, again, was the Americans' cartographic gaze, and it was fixed not just on the water, but on the people themselves—the Fijians, whom shipwrecked Americans might encounter, cast away on Fijian beaches, or whom Salem merchants enlisted to scramble over reefs in search of sea slugs. Like Emmons's description of the Tuamotuan a year earlier, Reynolds could not help couching his cultural observations in cartographic language. Like the reefs offshore, the Americans hoped to fix cannibalism into the expedition's survey and in doing so they sought to construct a safe environment for the expansion of American trade.[30]

Indeed, the imaginative transformation that ended with the finished hydrographic charts had already begun. Wilkes led a small band of officers and men to the top of Ovolau, a Fijian island home to a small Euro-American community. It also featured two-thousand foot vistas from which the Americans could envision the work ahead. "Desirous of fixing some of the main points in my own mind, as well as in that of the officers," Wilkes led his surveyors to the island's summit, which afforded panoptic views of the surrounding islands and reefs. From here, Wilkes and his men could imagine the charted ocean—to fix it in their minds, to borrow Wilkes's words. From atop Ovolau, the Americans placed themselves as the omnipotent surveyors looking down about the material environment, a view that they soon hoped to record for navigators who would look down, from nearly the same vantage, using new American charts. The reefs "could be traced for miles," Wilkes wrote, "and every danger that could in any way affect the safety of a vessel was as distinctly marked as though it had been already put upon our charts." With such perspective—with a clear vision of what lay ahead—the imaginative process of transforming wilderness had begun. From here, the marine environment in all its complexity seemed to appear on the Americans' mental charts in the precision of cartographic reality. The sea itself had become the chart. Wilkes's exercise in cartographic omniscience, at the very outset of the survey, served to connect the imagined process of dispelling wilderness with the actual practice of hydrographic method so central to the successful completion of the survey and the ideal of precision with which the Americans invested their charts.[31]

Thus envisioned, the Americans struck out across the islands in several subsurveys to frame Fijian waters as seas ripe for American commerce. Wilkes immediately brokered a commercial treaty with Tanoa, the most powerful chief in the islands, which regulated trade, assured protection for shipwrecked American seamen, and installed an American counsel to oversee trade and diplomatic relations with the United States. Here they constructed the diplomatic structure by which American trade and navigation, presupposed by the expedition's accurately fashioned charts, would hopefully flourish. The whole survey process was a complex orchestration—signal flags aflutter, boats tacking, and officers busily taking azimuths, barking orders, and jotting calculations. The Americans were certain that such attention to hydrographic practice assured precision and, by extension, the degree to which the chart could be trusted to faithfully reproduce the marine environment on paper. Reynolds sensed "a responsibility attached to the duty that forces one to be keen, watchful, and correct; if a Ship trusting to your chart gets into danger, who is to blame? We had much to do, knew nothing of the hidden shoals and the intricate passages that were to be found and laid down, and were limited to weeks to perform that which required months." The Fiji survey would test the Americans' discipline, their faith in cartographic method, and their own physical endurance. Not only did Fijian waters represent a complex and dangerous environment that, as Reynolds suggested, demanded unwavering physical and mental concentration, but the Americans operated under time constraints and worked close to shore where one eye also glanced warily into the palm-choked underbrush in anticipation of Fijian attack.[32]

Accurate hydrographic charts would turn the Fijian wilderness into a safe place for American commercial intercourse. With his boats off the island of Viti Levu, Emmons, for example, found one harbor "clear of shoals." He noted that it was "a good holding ground . . . well sheltered" with "ease of access and egress" and "an abundance of wood and water." In the *Official Narrative,* Wilkes boasted of "many . . . fine harbors" that "have never been visited." He intended the expedition's charts to call attention to so many "well-sheltered and commodious harbors." By the time the expedition departed the islands in August 1840, the Americans believed they had affected a cartographic transformation. The Fijis, remarked Lieutenant George Sinclair, were "as well surveyed as any group in the Pacific." They would

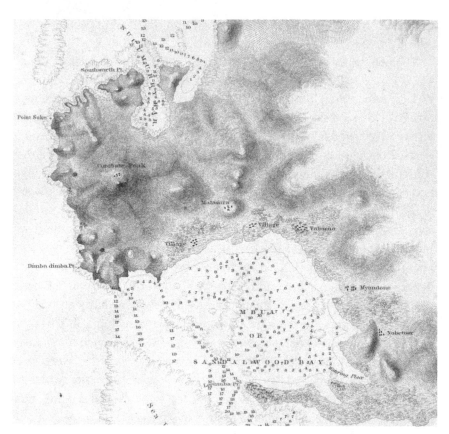

Detail of *Channel of the Southwestern Side of Vanua Levu* showing depth soundings and lines of triangulation around M'Bua, or Sandalwood Bay, in the Fijis. Courtesy of the Library of Congress, Geography and Map Division.

"hereafter find their *true position* on the charts," Emmons echoed. For Emmons and others, this American cartography was about truth—banishing mystery and conjecture in a faithful representation of the natural world. But, according to Emmons, it was not just the chart that better reflected the true reality of the physical environment; it was, as he said, the islands themselves that would find their own true position—on the chart, of course—but also in the minds of mariners. In this American cartography, it was not only existing charts that needed improvement, but the islands themselves whose geographic and hydrographic ambiguity now became fixed in place.[33]

Yet for all the Americans' sense of accomplishment, flaws in the survey emerged almost immediately that belied the cartographic ideal to which they aspired and, indeed, the extent to which the Fijis themselves could be incorporated into the American commercial world. As D. Graham Burnett has shown, it was the expedition's cartographic oversights, its surveying slips that, in part, embroiled Wilkes and his officers in a series of courts-martial, and professional accusations and counteraccusations that sullied the expedition's return to the United States. "The charts," Wilkes concluded in retrospect, were "the best encomium I can bestow on the united efforts of officers and men," yet even as the Ex. Ex. departed the Fijis, he ordered the schooner *Flying Fish* to return. "Captain Wilkes found that he had neglected fixing a certain island upon which much of the survey depended," Reynolds noted. "I hope and trust that Capt Wilkes will not claim too much for the chart," Lieutenant Sinclair conceded. It "might more justly be called a sketch than a survey. I would advise navigators to keep their eyes open when running in this group," he concluded, "even if they should have a cargo of charts aboard." Here, then, was the countervailing tendency of American hydrographic practice and, it might be said, the limits of American antebellum commercial empire more broadly. The Americans invested their charts with transformational power and their charts of the Fijis were unquestionably superior to earlier efforts, but they never achieved the precision and comprehensiveness they sought. Their cartographic ideal fell victim to time, but, more important, it was undermined by the agency of the marine environment and the agency of the Fijians themselves.[34]

Perhaps the greatest challenge to cartographic precision was the marine environment, which was vast, complex, and dynamic. "We were absent . . . ten days," wrote Sinclair of one subsurvey, "but were not able to do more than half the work that was allotted to us for the simple reason that it was more than twice as much as it was possible to do in that time." Faced with surveying a maze of reefs, he shrugged, "it would have required an age to fix them all and sound out at the same time." Emmons simply accepted the futility of Wilkes's orders. "By dark, our survey was completed—not indeed as I should wish—but as well as the time allowed would permit and well enough for all practicable purposes." Unlike terrestrial environments, where topographical maps might trace relatively static lines of territory or property, the paths of turnpikes, railroads, or canals, or the features of mountains peaks and passes, watery littoral environments lacked much semblance of linearity. Ocean waters were in constant flux, always changing the depth of water over reefs and sand bars even if by small degrees. Violent storms might substantially change coasts.[35]

Breakers and rolling swells, meanwhile, made scientific observation and measurement difficult, if not impossible. The ocean was no ideal laboratory for the practice of American science. Reynolds found himself, his boat crew, and his measurements continually undermined by the very environment he hoped to capture on the chart. "It was difficult in the extreme to make the observations," he groused, "the compass whirled like a top from the jumping motion of the boat . . . the seas that broke over us drenched all hands, and were sure to come as I was putting pencil to paper. We could scarce preserve our equilibrium on our seats—it is damnable." The midday sun left the brass sextants so hot as to be nearly unworkable by bare hands, and the men suffered from almost constant wont of food and sleep in the cramped boats. Reynolds complained of "baths of brine," and "a foothold that I could scarcely preserve from the depth I was in and the swash of the Seas." The marine environment was a precarious one for scientific work, undermining, with every oncoming swell, the methodical system that Wilkes had originally drawn up in his orders.[36]

The Fijians also contested American claims. They removed signal flags and, on one occasion, stole a surveying boat, which prompted Wilkes to burn a village in retaliation. It would not be the last. In June, a party led by Captain William Hudson, Wilkes's second in command, captured the Fijian chief Ro Veidovi, who had allegedly led the raid on the ill-fated bêche-de-mer trader *Charles Doggett* in 1834. He remained a prisoner for the remainder of the voyage and returned with the expedition to New York City in 1842, where he quickly died, probably of disease, a symbol of Fijian treachery avenged, of wildness now tamed, a specimen "collected, curated, and now rendered harmless by American naval power," as historian Antony Adler puts it.[37]

Emmons was probably too dismissive of the Fijians' own agency when his pilot and interpreter, a chief the Americans called Limbor, directed his surveyors away from the very island Wilkes had ordered them to chart. "The compass," Emmons remarked, was "too scientific an arrangement for Limbor's comprehension. So if I had followed the directions of King Limbor—who was sent with me as knowing all about these islands—I should finally have found myself a long way to leeward of the Asawa Group." The scuffle between Emmons and his pilot Limbor is characteristic of what cartographic scholar John Rennie Short has termed "cartographic encounters," in which indigenous and Western constructions of space commingled and were contested. Western ideas did not always prevail, at least at first. While it is impossible to know for certain, it seems unlikely that Limbor, a leader among people whose culture and means of interisland communication and war making was deeply rooted in indigenous navigational knowledge, would err so greatly in his own waters. Maybe there was a miscommunication with

Emmons, or perhaps Limbor was intentionally undermining the success of the survey, and, by extension, the American scientific claims to the islands, by leading them astray.[38]

Even as the survey progressed, the Americans continued to observe the Fijians for any sign of cannibalism, and on July 2, the men of the brig *Peacock* believed they witnessed irrefutable evidence. In Naloa Bay, north of Viti Levu, the Americans encountered two Fijian canoes. One contained a skull "yet warmed from the fire . . . and marked with the teeth of those who had eaten of it" and the other "some roasted flesh." To the men's disgust, one Fijian appeared to chew an eyeball. This "placed it beyond all doubt," Wilkes told his readers in the *Official Narrative*. Frederick Stuart, captain's clerk aboard the *Peacock*, agreed. "Just from what we have seen and heard, and secondly from the phrenological evidence," he concluded, "we have too strong proof." The naturalists quickly procured the remains, Wilkes remarked, for "a fathom of blue cloth." A fathom, or six feet, was thus both a measure of length commonly used in the survey's depth soundings and, here, of the value placed on the exchange between Fijians and the Americans hoping to procure these remains not only for scientific study, but for conclusive evidence of the supposedly savage practice. After the expedition's return to New York, the skull went on display at the new National Gallery in Washington, where many of the expedition's specimens resided next to the defleshed skull of Ro Veidovi. There, the skulls served "to remind visitors," historian Ann Fabian argues, "what those teeth had been used for."[39]

While Veidovi looked on in shackles, the Americans sought, and ultimately failed, to bring order to the Fijian wilderness. On July 23, three weeks after the discovery of the skull, Fijians killed two American officers, Lieutenant Joseph Underwood and Midshipman Wilkes Henry, who happened to be Wilkes's nephew, on the island of Malolo as they landed their surveying boat to procure provisions. After nearly three months of arduous surveying, it was apparent to the Americans that the islands would not be brought so easily into the American commercial world. The Americans had achieved neither comprehensiveness in their charting, nor order and control on the beach. At stake were differing and contested notions of the sea, who controlled it, and who could claim the material and human flotsam that occasionally washed up on Fijian shores. As Passed Midshipman George Colvocoresses noted, the Fijians considered anything drifted up on their beaches to be a gift from their gods. "All it contained," Colvocoresses noted, referring to wrecks and beachings, "is considered as belonging to the people of the district where the accident happens." The Fijians had their own cultural understandings of ownership, appropriation, and, not least, of who should control the waters surrounding their island homes. "Savage nations," Secretary of the Navy Paulding had warned Wilkes, were "unacquainted with" or possessed "but vague ideas of the rights of property." To the Americans, this was one of the defining characteristics of Fijian savagery. The Americans dismissed such notions if they cared to think much about them at all. They proceeded to survey and chart the marine environment regardless of contrary or competing understandings. Indeed, cartographic science assumed that uncharted waters were little more than blank spaces on which explorers and surveyors could impose their own meanings, dispelling or displacing others. Yet the "Massacre at Malolo," as the Americans came to call it, suggested that Fijians continued to undermine American scientific, commercial, and, by extension, military pretensions. What looked like an ordered commercial space on the chart revealed itself to be a contested environment in reality.[40]

The Americans set out to punish the Fijians of Malolo, bringing the military power of their naval expedition to bear and creating the kind of order that had apparently eluded their survey. Environmental historian Lisa Brady has argued that the Union Army's destruction of the Southern landscape during the American Civil War was a deliberate strategy to defeat the Confederacy by turning improved land into a wilderness. Here, however, the opposite was at work. Wilkes and his men intended to tame Fiji, its people, and its culture by razing wilderness. If the peaceful pretenses of science and the hydrographic survey could not accomplish it, Wilkes could summon the military power of his naval force, drawing on the Navy's record of violence against Pacific islanders in reprisal. Command of the landing force fell to Lieutenant Cadwalader Ringgold with directions from Wilkes that "every man or native capable of using a club, or stone is to be destroyed." On July 24, Ringgold set out with three divisions of seamen and Marines in a scorched earth march across Malolo, but not before the schooner *Flying Fish*, which Wilkes ordered to cover the landing from the water, grounded on an uncharted reef. The Americans pressed on, setting fire to coconut and yam fields and flattening two villages. At the village of Sualib, the Fijians mustered a defense behind ditches and palisades, but the assault of American musketry and Congreve Rockets proved overwhelming, and they fled into the jungle. At day's end, nearly one hundred Fijians had perished; one American was injured.[41]

The expedition had reduced the island to smoldering ruin. "Everything contained within the walls was utterly destroyed," remarked Silas Holmes, "it being the object . . . to make the island desolate." The next day, when a group of Fijians issued from the jungle on their hands and knees to meet the Americans, they "begged pardon, supplicating forgiveness, and pledging that they would never do the like again to a white man." Wilkes was satisfied. "The punishment," he later testified, "was sufficient and effectual." Not surprisingly, he cited both the safety of "our countrymen on their adventurous voyages" and the Fijians' "horrid appetite for cannibal repasts" as justification. Wilkes was convinced that he had brought order to the sea. "Such has been its effect on the people of Malolo," Wilkes boasted about his reprisal for the deaths of Underwood and Henry, "that they have since been the most civil, harmless, and well-disposed natives of the group." Through use of force, the Americans sought to achieve the order and safety that had alluded them during the three-month Fiji survey, which had been continuously undermined, first, by the vastness and agency of the marine environment and, second, by the Fijians who contested American claims by acts of resistance large and small. When the Americans left Fiji burning behind them, they had chastised an entire island, but they also left chastened themselves.[42]

Commerce, science, violence—these were the ingredients of American commercial empire in the Pacific during the antebellum era. All three came together in the voyage of the Ex. Ex. When the expedition returned to New York in the summer of 1842, it had transformed the way Americans thought about the Pacific Ocean at the beginning of a decade marked by American Manifest Destiny and westward expansion across the continent. The American public and American writers such as Cooper and Melville pored over the expedition's five-volume *Official Narrative*, which Wilkes published in 1844, drawing on the journals of his officers. "The chapter on currents and whaling grounds, is, of itself, a full return for all the expenses incurred by the government in the voyage," wrote *Hunt's Merchants' Magazine* in 1845. Together, the expedition's charts and texts helped to usher a transformation in American sea literature away from the Romantic visions of ocean adventuring so prevalent in Cooper's earlier novels, toward greater fidelity to what the Americans had apparently seen, observed, witnessed, and charted. For an American readership, these writers used the expedition's charts and texts to reframe the public's understanding of the oceanic world and American mariners' place on it. As literary scholar Thomas Philbrick and others have shown, Cooper, in some cases, lifted entire passages of Antarctic voyaging from Wilkes, nearly verbatim. Philbrick's son, the historian Nathaniel Philbrick, argues that "the pages of the *Narrative* provided a visual link with the exotic world of the South Pacific . . . that no other American book could match."[43]

Vision, of course, was also the aim of the expedition's cartography, which expanded the ways mariners could see, understand, and avoid navigational dangers, guiding them safely to protected anchorages. There were 180 of them, published in a two-volume *Hydrographical Atlas* and as individual charts issued by the new Naval Observatory and Hydrographical Office in Washington. From Antarctica to the Fijis, the expedition's charts both reflected and hastened a growing American commercial presence at sea. These were the products of some 250 individual surveys, countless soundings and azimuths taken, and, above all, a nearly unshakeable faith in the precision of modern cartographic method rooted in the trigonometric survey. The finished charts were impressive testaments to voyaging in little-known seas; they were, in many cases, vast improvements over preexisting cartographic and hydrographic knowledge. On them rested the Americans prerogatives to the discovery of a new continent. On them also rested American claims to the water boundary between the Oregon Territory and British Columbia. After the expedition's exploration of the Pacific Northwest in 1841, its charting of the Columbia River bar, and the Straits of San Juan de Fuca, argues diplomatic historian Frederick Merk, the British were forced largely to concede their position south of the forty-ninth parallel. "I send for your inspection a traced copy . . . of Wilkes' Chart of the Straits of Juan de Fuca, Puget's Sound, &c., &c.," Secretary of the Navy George Bancroft wrote to British foreign secretary Lord Palmerston during the boundary negotiations in 1848, two years after the signing of the Oregon Treaty establishing the forty-ninth parallel as the western boundary between the United States and British Canada. The expedition's charts and its reports, Merck argued, "came with new emphasis to a community that had expanded its horizons and enlarged its desires on the Pacific." When the Americans sought to make claims vis-à-vis the British, they turned to the exploring expedition's precedence and the chart's authority in matters of diplomacy and westward expansion by land and sea.[44]

The expedition's charts fixed the Pacific in an American cartography. Nowhere is this more evident than in the names that now dotted that vast ocean. Christian Jacob has written that toponymy, or the cartographic process of naming places, is "a mode of symbolic appropriation that provides virgin territories with a memory, a grid that dispossesses space of its otherness and turns it into an object of discourse." The name "is thus a signature, a claim of precedence and of symbolic ownership," the ultimate aim of which, he argues, is "political and colonial mastery." The charts marked Disappointment Bay, in Antarctica, bestowed by Wilkes because, he said, "it seemed to put an end to all our hopes of further progress south." Emmons remarked that Fiji's Peacock Harbor was "a compliment to my floating Home—for which I feel a singular attachment." The Americans named other places for shipmates and sweethearts, so that seemingly all had their place in the expedition's cartographic wake. Most significantly, Wilkes Land ultimately came to span the fifteen hundred miles of Antarctic continent set down on the expedition's chart despite the disputations of rival European explorers. It has stuck, unlike many other names whose resonance has faded over time, replaced by new meanings and new cartographic, imperial, and national authorities.[45]

The Americans used toponymy for remembrance as well. The Fiji chart marked Henry Island in Underwood's Group in memory of the two slain American officers. There is perhaps a marked possessiveness in the latter that went beyond mere cartographic convention to represent the actual place where the Americans' bodies now rested. In their deaths and on the chart, they had literally claimed the islands for themselves. Even as the Americans groomed the sand over the graves of their shipmates to hide them from prying cannibals, the chart clearly identified their last resting place. "We trust no future hydrographer will venture to remove these touching landmarks," wrote one reviewer of Wilkes's *Narrative*. The waters off Malolo, Reynolds remarked, took the name Murderer's Bay, which "served as a sad memorial of their own loss, of the disposition of the people, and as a warning to future adventurers." Naming places is the prerogative of the cartographer-explorer. It is an act of control and appropriation. As the trigonometric survey filled in blank stretches of sea in numbered depth soundings and triangular baselines, these names inscribed meaning, narrative, and a history to waters the Americans sought to dominate primarily through commerce, but, if need be, through naval power as well.[46]

Still, American claims to control the natural world and the savages that seemed to inhabit an ocean wilderness remained contested—in places like Fiji—by the marine environment and by the indigenous people themselves. Despite Wilkes's claims about the effective use of force to compel the Fijians into submission, the United States sent American warships to Fiji in 1855 and again in 1858 to settle depredations to American commerce. While vastly improving navigation in Fijian waters, the expedition's charts did not secure them. American vessels still wrecked. In June 1849, nearly a decade after the Ex. Ex. left the Fijis, William Lloyd Garrison's *Liberator* noted the loss of the Nantucket whaleship *United States*, "struck on a reef in the vicinity of the Fee Jee Islands." By the paper's account, shards of splintered timbers washed ashore marked "United States." The fate of the vessel seems an apt metaphor for aspirations—scientific and commercial—in Pacific waters where American pretensions, in many cases, went to pieces among the reefs and shoals of a dangerous marine environment. Yet, as Methodist missionaries Thomas Williams and James Calvert noted in 1859, the "great utility" of Fiji's reefs was "certain. The danger caused by their existence will diminish in proportion to their position and outline become better known by more accurate and minute survey than has yet been made. To the navigator possessing such exact information, these far-stretching ridges of rock become vast breakwaters, within the shelter of which he is sure to find a safe harbour. . . . In many cases a perfect dock is thus found." To Williams and Calvert, Fiji's marine environment was in the process of an imagined transformation that could only be effected by more accurate charts and continued surveys, in which the reefs might be turned from fearsome navigational hazards to breakwaters that, rather than dangers themselves, would provide protection from the ocean's destructive natural forces. Clearly, the Ex. Ex. had not quite achieved this, but its survey and charts marked an important moment in which this process of recasting the natural environment to suit American commercial interests commenced.[47]

Intent on asserting the control that eluded the Americans in the survey and having lost officers to Fijian attack, Wilkes had turned to the last dominant card in his hand—the use of force—which proved such a controversial and excessive application of military power that a court-martial took up the question—among many others—following the expedition's return to the United States. Military historians do not often think of environmental destruction beyond the collateral damage of battle, but in the Fijis, the use of military force was both the product and outcome of environmental considerations that, in large part, derived from a sense that military power—the application of overwhelming violence—could compel the Fijians and extend some measure of control over a natural world that the Americans had first sought through hydrographic surveying, chartmaking, and scientific inquiry more broadly. Indeed, in the Ex. Ex., the twin instruments of science and violence were complementary . Not only were environmental considerations crucial in achieving military, scientific, and national aims, the use of naval power was an important way to bring oceanic environments—which included indigenous people—into the realm of American commercial empire.

The voyage of the Ex. Ex. marked an important moment in American visions of the Pacific Ocean and the expansion of commercial empire across its waters, among its reefs, and onto the beach where the work of trade and cultural encounter beckoned and threatened American prospects. The U.S. Navy's role in protecting these traders, their ships, and cargo and in seeking reprisal for the destruction of it, proved an important part of the navy's work in the antebellum era. Less understood is the role of surveying and cartography in extending this empire of commerce and science. In bringing order to the Pacific, staking claims of discovery to the Antarctic and the "savage coasts" of Pacific islands and atolls through its charts and, in some cases, through destructive violence, the expedition reframed a marine environment that the Americans aspired to claim, if not as American territory, then as supposedly safe for navigation, trade, and cultural encounter. In so many American names and narratives, the charts, as Wilkes liked to put it, spoke for themselves as powerful testaments to Americans' intentions and precedence, but also as a deeply flawed representation of an oceanic environment more difficult to control than the finished charts let on. "When I was a boy . . . the name of Wilkes, the explorer, was in everybody's mouth," remembered Mark Twain, whose own pen name he derived from the shallow river soundings of his native Mississippi River. "What a noise it made, and how wonderful the glory! . . . Wilkes had discovered a new world and was another Columbus."[48]

# 3 The Common Highway

. . . . . . . . . . . . . . . . . . . . . . . . . . . . . . . . . . . . . . . . . . . . . .

It might seem an absurdly hopeless task thus to seek out one solitary
creature in the unhooped oceans of this planet. But not so did it seem
to Ahab, who knew the sets of the tides and currents; and thereby . . .
could arrive at reasonable surmises, approaching almost to certainties,
concerning the timeliest day to be upon this or that ground in search of
his prey.

—Herman Melville, "The Chart," in *Moby-Dick*

"How the skipper saw the crowded and rudderless wreck of the steamship,
and Death chasing it up and down the storm," wrote Walt Whitman in his
paean to Heroes in the poem "Song of Myself." He was referring to the loss
of the steamship S.S. *San Francisco* bound from New York on its maiden voyage to its namesake port when it encountered a storm while traversing the
Gulf Stream three hundred miles from Sandy Hook. *San Francisco* had been
built for the Pacific Mail Steamship Company and launched the previous
June, promising to be "superior to any thing that has yet made its appearance" and "the finest steamship on the Pacific," predicted the newspaper
*Daily Alta California*. It was to be a marvel of modern mechanized technology, safety, and luxury designed to make speedy passages between the isthmus of Panama and the Golden Gate, and thereby usher a great migration
of Americans westward, not by land, but by sea. On Christmas Eve, 1853,
the steamer "experienced a most terrific gale from the northwest," Captain
James T. Watkins reported, "which continued to increase with great violence until it blew a perfect hurricane, with a very high, irregular sea."
Waves carried away the steamer's smokestacks. Seas swept its decks. A
giant wave rolled over the spar deck, taking with it the upper saloon and
one hundred souls. The ship lost power. Attended by a number of nearby
vessels, which could not get near enough in the high seas to rescue all the
ship's passengers and crew, it tossed at the mercy of wind and waves until
January 6, when it finally sank with the loss of more than two hundred of
its seven hundred passengers and crew. Ultimately, approximately five hundred survived thanks to the determined efforts of those nearby vessels—

the subject of Whitman's lyricism—standing by to rescue the helpless passengers. Whitman, in some respects like Cooper, Thoreau, and Melville, conjured a sea whose wildness could humble human pretensions, but at the same time serve as a stage for great feats of human endurance and bravery.[1]

As *San Francisco* drifted into 1854, a maritime crisis slowly unfolding, the secretary of the navy summoned Lieutenant Matthew Fontaine Maury, asking him to predict the location of the stricken steamer so that two rescue vessels might be dispatched. As the superintendent of the Naval Observatory and Hydrographical Office, Maury knew more about the North Atlantic's winds and currents than perhaps anyone. But how to find a small ship in a big ocean? "A chart was prepared," Maury recalled, "to show the course of the Gulf Stream at that season of the year." He drew lines "to define the limits of her drift." Though the vessels arrived too late, Maury's prediction of *San Francisco*'s drift based on his knowledge of wind and current direction, speed, and velocity and the physics of what he referred to as "cyclones" proved nearly precise. In his seminal book *The Physical Geography of the Sea*, published in 1855, Maury used the example of the *San Francisco* as evidence of his broader philosophy that the marine environment obeyed universal laws that he and his staff at the observatory were in the process of divining through "a system of philosophical deduction . . . on shore" and based on the thousands of ship logs that, by the 1850s, were flooding to Maury from all seas of the world. Mariners needed only to heed these laws, which Maury set within a new kind of cartography—his *Wind and Current Charts*—to harness a watery nature, which by Maury's work had taken on some degree of predictability. The mariner could make "reasonable surmises, almost approaching certainties," Herman Melville wrote in *Moby-Dick*, which referenced Maury's work at the observatory in Ahab's quest to find his whale.[2]

Thus, in broadly analogous ways, Maury was doing for the deep, open ocean what Wilkes and the Ex. Ex. had done for much more localized coastlines in places like the Fijis, that is, using hydrography and the chart more specifically to bring rationality to a marine environment now traversed by an oceanic commercial fleet of merchant vessels and whalers as the American maritime world reached its peak in numbers and economic value during the antebellum era. Maury's charts and sailing directions played a significant role in the making of American commercial power in the decade of the 1850s, transforming the sea into a racetrack or common highway in which sleek clipper ships—some with trailblazing names such as *Daniel Boone*—competed for the best time from New York to San Francisco, Maury's charts in hand, while packets plied the North Atlantic between New York, Liverpool, and Le Havre on strict timetables. Of course, this did not always lessen or eliminate the sea's dangers. Modern, mechanized vessels such as *San Francisco* still foundered; Maury often struggled to win the eyes and ears of navigators, and he grappled with the chart's limitations to portray the marine environment in ways that both represented the many complexities in the movements of wind, water, whales, and other natural phenomena, and at the same time, made these same forces immediately discernible.

Maury was a controversial person in his time, and historians have continued to debate his place in American science and naval affairs, but in many ways, he was a unique figure whose significance can best be understood, first, through his charts and his system of data gathering, which revolutionized the way mariners viewed, understood, and used the oceanic environment and, second, through his ability to bridge three increasingly divergent communities, all of whom had a stake in knowing the antebellum ocean. His efforts can be more fully understood not as a scientist alone, nor as a naval officer, nor as a benefactor of the maritime community, but in the intersection of all three. In these entangled identities and in the different masters he served, he became an iconoclast to each. Yet Maury alone was able to mediate between the navy, civilian science, and the maritime community at a time when each often looked askance at the others. In doing so, he was a central maker of an American commercial empire in which scientific inquiry, naval affairs, commercial voyaging, and the agency of the natural environment converged and intermingled.[3]

As a young passed midshipman, Maury met Cape Horn not with fear, nor with the bold adventuring spirit of earlier navigators like David Porter, but with a scientific eye to discern patterns of wind and water that combined to make one of the most fearsome environments of the maritime world. In 1831, he reported to the sloop-of-war *Falmouth* for a voyage from New York City around Cape Horn to the Pacific. As sailing master, he was responsible for the ship's navigation. In preparation for the forthcoming voyage, he consulted charts, sailing directions, and various voyage narratives, hoping to glean useful, practical knowledge, but found the information scattered, fragmentary, unsystematic, and often contradictory. So he resolved to write about it.

In July 1834, having returned from the Pacific, he published an article in the *American Journal of Science and Arts* on the subject of navigating Cape Horn in which he juxtaposed the ocean wilderness navigated by the experienced mariner with an ordered ocean derived from empiricism and rational thought. Maury began by conceding Cape Horn's terrors. "The most robust constitutions," he wrote, "overcome by long exposure to it, succumb to its severity; they may bear up against it for days, but the hardiest crew, exhausted at last by incessant toil, are forced in despair to give up the ship, clogged with ice and snow, to the mercies of the contending climates." Maury framed marine navigation as a trial of man against a wild environment in which the natural world often triumphed. But he did not subscribe to this way of thinking. His mind was more analytical, impressed by the possibilities of scientific inquiry and less governed by fear and uncertainty. "Under the guidance of certain circumstances," by which he meant a more systematic knowledge of the Horn's wind patterns, "the navigator may be greatly assisted in conducting his vessel in safety through the tempestuous sea connecting the Pacific with the Atlantic." Based on his own experience and that of various ships' logs he consulted, he urged fellow navigators to continue southward until they found favorable winds, which seemed to privilege inshore and offshore passages at different seasons of the year. Rather than accepting the dangers of a seemingly tumultuous and irrational environment, Maury urged scientists and mariners to interrogate the natural world to find the patterns of order that, he was certain, would make rounding Cape Horn a less terrifying—and less deadly—experience.[4]

Maury hoped not just to render the environment more clearly to the navigator, but to make science less inscrutable to the mariner and the naval officer. In 1836, he published a book titled *A New Theoretical and Practical Treatise on Navigation*, which did not break new ground so much as it recast existing navigational science in a graspable way. Before the navy established its academy at Annapolis in 1845, the education of naval officers resided primarily aboard ship, directed by the captain and his lieutenants. Young midshipmen needed to be proficient in navigation. A basic understanding of it was essential for any officer serving aboard ship. But instruction at sea demanded utility, not cumbersome tomes. Maury stated his rationale in the preface to the first edition. "It is not pretended that new theories are set forth . . . but it is believed that those which have already been established, are here embodied in such a form, that the means of becoming a theoretical as well as a practical navigator, are placed within the reach of every student." Maury strove for clarity, keeping in mind that his readers, though all young gentlemen, would come to the navy with various levels of proficiency. Maury wrote that the work should be "an elementary one, adapted to the capacity of all." It was a democratic text, assuming no prior knowledge above basic arithmetic, but promising to teach all the mathematical principles behind the science of navigation unencumbered by the opacity that increasingly rendered science the exclusive realm of the intellectual.[5]

Maury hoped that his manual would professionalize the act of navigation within the naval officer corps while bringing science more firmly into a naval service whose ranks remained largely opposed or indifferent to the practice. Maury's early career coincided with a time of intense professional anxiety and promotional stagnation in the naval officer corps, and so he quickly became one in a generation of young reform-minded officers who adopted progressive views about the roles of science and technology in the service. Among other things, Maury saw science as one way to advance through the profession during times of peace, constriction, or otherwise slow promotion. The book, Maury wrote, would be "the first nautical work of science that has ever come from the pen of a Navy officer, and upon its merits I intend to base a claim for promotion." In particular, Maury chafed at the system of officer education that privileged memorization of navigational practices more than the underlying mathematical principles behind them. By the 1830s, he was also calling for reform of the navy's administrative structure and its system of promotion.[6]

His book met critical acclaim in and outside of the navy, and the service soon adopted it for the instruction of its midshipmen. One officer thought its "explanations of the principles . . . both ample, easy and well-arranged." The Naval Lyceum, an early center of naval intellectualism, praised it for "a simplicity that has heretofore been generally wanting in books on Navigation." But the most revealing review came from the pen of Professor A. G. Pendleton, a naval instructor of mathematics. The book was "best calculated," he thought, "to induce a love for the prosecution of the study of navigation as a science, and not merely as an art." Maury's *Navigation*, though not the first text to do so, encouraged navigators to think about the sea scientifically rather than strictly by the hard-won lessons of experience, the folkloric meanings rooted in the maritime world, and the publications emanating from the maritime community. Its simplicity, meanwhile, appealed to mariners and aspiring naval officers, prefacing qualities that would win Maury the slow acceptance and respect of mariners and, not coincidentally, the suspicion of many in the American scientific community.[7]

Praise for Maury's manual proved enough to mark him as one of the most accomplished scientifically minded officers in the navy. The list was not long, particularly in the 1830s before the voyage of the Ex. Ex. initiated a squadron's-worth of young officers in the intricacies of the trigonometric survey. Even Joseph Henry, a professor of natural philosophy at the College of New Jersey and future Smithsonian secretary, who by the 1850s would become one of Maury's chief critics, wrote to inquire about this "young man" who, he had been told, exhibited "great industry, ability, and promise." In June 1842, after convalescence from a stagecoach accident that left him without the full use of his right leg and thus practically unable to go to sea again, he received orders to report to Washington as the new superintendent of the Depot of Charts and Instruments. Maury inherited command of the depot from Wilkes's successor, Lieutenant James M. Gillis, a noted astronomer who had advanced the depot's research in celestial science, and initially, Maury seems to have continued Gillis's emphasis on the heavens in addition to the more prosaic, but no less important, duties of rating the navy's chronometers and supervising the dissemination and collection of charts and other navigational instruments to naval vessels.[8]

Maury had initially sought a position in the United States Exploring Expedition, but resigned his name when its command fell to Wilkes, whom Maury despised. To Maury, Wilkes was "this favorite of imbecility" and "the only officer in the Navy with whom I would not cooperate." The quarrel seems to have originated in what Maury deemed Wilkes's inept preparations in procuring instruments for the expedition in Europe and Wilkes's maneuvering to secure command of the expedition. Historian D. Graham Burnett has gone so far as to suggest that Maury's whole cartographic program at the observatory was motivated, at least in part, by competition with Wilkes and his expedition. "How to rival Wilkes and the U.S. Ex. Ex. without ever leaving a desk job in Washington? The answer lay in maps," Burnett contends. In the 1830s and 1840s, science was already a fairly small corner of the naval officer's profession. The barely concealed distrust between the navy's two most important antebellum scientists illustrates the fragmented and sometimes internecine nature of scientific work within the service.[9]

But if Maury expressed disappointment over Wilkes's command of the Ex. Ex., he nevertheless assumed the duties of the navy's chief scientist in what would soon become a world-class observatory, making Washington, D.C., and the navy suddenly one of the premier places in the world to practice science. When the navy established the Naval Observatory and Hydrographical Office in 1844 as a successor to the Depot, the nation's capital joined Paris, Greenwich, Cincinnati, and Cambridge, Massachusetts, as one of the few places with the capacity for serious astronomic observation. It fell to Maury to usher naval science through this transition from the depot to the observatory's new home on 23rd Street overlooking the tidal flats of the Potomac River, where mists, miasmas, and mosquitoes often obscured the night sky and nuisanced the observatory's staff, which by the 1850s numbered as many as twenty officers and civilians.

In the process of going through the old logbooks of voyages made by naval vessels then gathering dust in the depot, Maury stumbled upon a rich archive of data bearing on the marine environment, the very sort he had sought, unsuccessfully, as he looked forward to *Falmouth*'s voyage eleven years earlier. Kept daily as a requirement of the service, the ship's log represented a window into marine environments all over the world, documenting not only the daily goings-on aboard ship, but also navigational and meteorological information, course headings, and any other noteworthy happening that occurred aboard ship during a voyage, including storms, barometric pressure, wind, and current speed and direction. The navy's logbooks seemed an untapped resource. So Maury set about "to overhaul the old logbooks," looking entry by entry for clues to the workings of the sea and the atmosphere. He was curious about winds and currents, natural forces that could most accelerate or impede voyages under sail. In combing through logs, Maury cast the same eye for systematic examination that he had demonstrated in his article on the navigation of Cape Horn. "By comparing and discussing these observations," Maury concluded, "information . . . valuable to the commerce of the country might be elicited." But the data gleaned from the logs was often crude. Other than the fundamental observations of latitude, wind direction and velocity, and sea state, logs had no standardized structure. Each log was as different as its keeper or ship's captain, who as yet had no expectation of contributing them to some broader program of scientific inquiry. Knowledge about the marine environment had been collected "without system," Maury complained, "and with little or no regard to the facts, which I wish to obtain from them." Without a more structured approach to data collection, in other words, he could not draw firm conclusions of practical value.[10]

Not content with the logs alone, Maury sought cooperation from the navy and the American merchant marine, but through much of the 1840s, the response was cool. In December 1842, he issued a circular to commanding officers in the navy as well as commercial ship owners and masters through his administrative superior in the Bureau of Ordnance and Hydrography. He requested, especially, "all that valuable information relating to the navigation of distant seas." In particular, he sought observations on winds, currents, tides, weather, magnetic variation, and *vigias*—a term for rumored but still unproven hazards to navigation that dotted so many nineteenth-century charts. Maury reached out to naval and merchant mariners who knew well the pressing need for more accurate charts and the practical benefits to be gained from them. In the circular, he promised to make the information he obtained "accessible to navigators" and proposed to "open a regular channel of communication" with them. Here, Maury realized, lay the potential for a novel and mutually beneficial relationship that could take him to the remotest waters without, as Burnett has suggested, leaving the confines of the observatory. His excitement was palpable. "How pregnant and full of meaning would be the spectacle of a floating Observatory in every man of war," Maury exclaimed at the end of 1847. If Maury could not go to sea himself, he could enlist naval and commercial fleets as his observers on the water. He had few qualms about welcoming contributors into his scientific program. His hope, beginning with his manual on navigation and continuing through his superintendence of the observatory was to make naval officers and merchant captains into scientific observers, or at least to fashion them into important participants in the scientific process.[11]

At first, Maury's requests were met with indifference or skepticism in the navy and the broader American maritime world. In the mid-1840s, the union between science and the navy remained an awkward and mostly unrealized one. Despite the progressiveness of some officers such as Maury, Wilkes, John Dahlgren, Matthew C. Perry, and Charles Henry Davis toward matters of science and technology, the U.S. Navy generally remained a conservative institution administered—Maury would say lorded over—until 1842 by the Board of Navy Commissioners, a three-person board of senior naval officers who reported to the secretary of the navy. With no clear promotional incentive to seek scientific duty or to carry out specialized scientific work, American officers could not be convinced to embrace a program of research at sea that required rigorous measurements and observations. As much as it excited Maury, not all or many officers cared to make their warships into floating observatories. Through the decade, Maury did receive reports from "a number of commanders" who had "of their own accord, entered heartily into the subject." But he could not compel their participation, and the Navy Department did not order it. Diplomacy, war with Mexico, and the innumerable exigencies of duty on distant stations took precedence over hydrographic science. In order for these observations "to tell well," Maury knew, "every vessel should be an observer and contributor."[12]

Maury also grew frustrated by the lukewarm response from the maritime community, which seemed to stand so much to gain from his charts, and he was convinced that these shipmasters and navigators were a tradition-bound lot, suspicious of navigational knowledge penned by a land-bound naval officer rather than the marine societies and privately run publication houses that had long been the repositories for charts, sailing directions, almanacs, and coast pilots. Among these seamen, the circular "was not regarded. . . . No response whatever was elicited, and the appeal passed by unnoticed," Maury grumbled. The basis for this indifference, he surmised, was the traditional belief system of the mariner in which sailing routes were determined and revised by experience, not through methodical investigation. "It is hard to get old sailors out of old notions," Maury groused in a letter to Congressman Julius Rockwell. "Two vessels sail together for the same place," he hypothesized, "one arrives two, three, or even twenty days before the other, according to circumstances. This is called 'luck,'" he continued, "and the master who makes short passages is called 'a lucky fellow.'" Of course, Maury placed no credence in luck. He attributed speedy voyages to natural "laws" and the "order of nature." The mariner's intransigence frustrated him. His pen sometimes lashed out at "pig-headed" captains who seemed "unwilling to learn, especially from one who has never performed the voyage." Merchant captains, it seems, were loath to cast aside their experiential and folkloric understandings, especially for the advice of a land-bound naval lieutenant with one good leg and strange advice about making voyages other than by long-established routes. By 1847, Maury was convinced that he needed to win these mariners to his view. "The object," he concluded, had never been "presented in the right way." In winning the respect of merchant captains, Maury would need to demonstrate the usefulness of his charts and to do so in ways that were plainly evident to the mariner. He would need a new kind of chart.[13]

Where in 1842 his polite circular had fallen largely on deaf ears, Maury now believed that the chart carried a measure of authority and—in the ideal—immediate obeisance to the cartographer's supposed omniscience. Certainly, this was the case with Maury's "Fair Way to Rio," which formed the first of what was to become his famous *Wind and Current Charts* series. Maury understood the visual power and practical potential of the nautical chart, and so, using his meager sources, he began work on a track chart of the North Atlantic, intending to show a new and faster route from the United States to Rio de Janeiro. Throughout the mid-1840s, he had steadily collected logs, not only from the navy but also those gleaned by a few cooperative former sea captains and their associates in Boston, Salem, New Bedford, and Nantucket. In looking at the passage to Rio de Janeiro, a busy destination for American grain shippers, Maury found that captains followed a circuitous zigzagged path, "crossing the Atlantic twice, or nearly twice" to double Cape St. Roque, the shoulder of South America thrust far into the Atlantic. Based on his study of ship's logs, Maury advocated a much closer adherence to the coast, where favorable winds and currents would prove the mariner's ally. He drew up a chart with a giant pointer finder oriented southward as if to stiffen the resolve of the wary navigator. Maury's cartographic hand made for awkward symbology, but it nevertheless suggested his emerging understanding of the chart's visual power to convert the maritime world to his cause.[14]

The track chart, which Maury named the "Fair Way to Rio," shaved days off the old voyage and gave him the authority to begin a comprehensive investigation of winds and currents. In January 1848, the bark *W.H.D.C. Wright* departed Baltimore with a cargo of grain bound for Rio. It arrived the next month after a passage of thirty-eight days using Maury's new route laid down on the chart. The voyage of "the alphabetical barque," as Maury and his staff gleefully dubbed the *Wright*, had saved seventeen days over the usual passage and proved, in Maury's words, "the first fruit of the Wind and Current Charts." A powerful endorsement of Maury's efforts, *Wright*'s voyage began to break down the entrenched belief system of so many mariners, which Maury had previously found so unwavering and pervasive. "Navigators now appeared for the first time to comprehend clearly what it was I wanted them to do, and why," Maury remarked. "They appreciated the importance of the undertaking, and came forward readily with offers of hearty, zealous, and gratuitous co-operation." Maury had, in a moment, captured the attention of the maritime world by appealing to its purse and by making scientific principles, in his words, "as clear . . . as the Sun at midday in a cloudless sky." Such analogies, which appeared beginning in 1851 as his *Explanations and Sailing Directions to Accompany the Wind and Current Charts*—a sort of textual corollary to the chart—were not lost on the mariners who read them with growing interest.[15]

Even in his earliest writings, Maury had displayed an elegance in his prose and a familiarity with literary conventions that by the 1850s would serve him well as a writer, a scientist, and a popularizer of scientific ideas. He had the ability to deftly construct images with the written word, and later, of course, with his charts, too, a characteristic of the Humboldtian scientific encounter with the natural environment that not only influenced scientists such as Maury but American writers such as Henry David Thoreau, Ralph Waldo Emerson, and Herman Melville. His description of the

Gulf Stream in *The Physical Geography of the Sea*, for example, began famously with the lyrical line, "there is a river in the ocean." And if his writing evoked imagery, his charts could be equally potent tools of narrative. Like the voyage of the Ex. Ex., stories and cartographies were deeply intertwined.[16]

Maury believed that hydrography and ocean science more broadly promised to transform the ways mariners, scientists, and naval officers viewed the sea. He likened the current state of navigational and hydrographic knowledge to "any Indian trail through the wilderness." Such routes were "well-beaten," but "most curious and crooked." Maury hoped to replace voyaging along these experiential paths with the ordered linearity of the chart so that the mariner seemed no longer the supposed savage, but instead the frontiersman trailblazer wielding an enlightened science to beat a new path "not . . . upon trees, as in the wilderness, but upon the wings of the wind." In the era of Manifest Destiny, such analogies were not just lyrical. He went further. "All the rest of the ocean . . . was blank, and seemed as untraveled and as much out of the way of the haunts of civilized man as are the solitudes of the wilderness that lie broad off from the emigrants' trail to Oregon." As crooked indigenous trails supposedly gave way to national turnpikes, so too did the paths of the sea emerge through the efforts of an increasingly national scientific program whose aims, in part, were to rationalize nature and thus expand the economic and cultural reach of the United States in the process. When Maury invoked an ocean wilderness, he did so not to suggest that the sea was inherently wild, terrifying, or chaotic. "Nature is perfect," he had told an audience of the Geological and Mineralogical Society of Fredericksburg, Virginia, as early as May 1836, "her laws are universal and operate every where alike." Rather, he deployed this convention to establish the central allegorical function of his hydrographic and cartographic work. That is, the sea seemed wild and untamable so long as it remained uncharted or ill-charted. "The pioneer goes and returns: 'Which way did you go?'" Maury wrote of this hypothetical navigator's frontiersman in the preface to *The Physical Geography of the Sea*. "How lies the route? Give us your sailing directions." To chart the ocean would be to civilize it. Mariners thus joined Oregon emigrants as antebellum pioneers of empire who transcended the West to encompass the seas beyond.[17]

From his first work on the winds and currents of Cape Horn, Maury had sought to rationalize the ocean environment, to interrogate it, and then represent it systematically in texts and charts devising a method that transcended the folkloric and rule-of-thumb navigating that he thought so wasteful, inefficient, and often dangerous. Flush with the example of the *Wright*'s voyage to Rio, Maury now set out to convince the rest of the maritime world of the value of hydrographic work by integrating mariners themselves into the scientific process as observers of the marine environment. By 1844, he had fashioned an abstract log, which he distributed to shipmasters, naval captains, shipowners, insurers, and merchants, which systematized and codified the practice of scientific observation at sea. Across the top, Maury constructed a series of columns, beseeching mariners to fill in information about wind direction and velocity, the direction and speed of currents, ocean temperatures, barometric pressure, whale sightings, and any other observations that seemed pertinent to note. Along the left side of the log, he drew hourly intervals at which time mariners should take their readings. To express the intangibles of weather, Maury urged the use of abbreviations and symbols—"f" meaning fog, "s" for snow, "g" to indicate gloomy, dark weather, and so on. "By the combination of these letters," Maury remarked, "all the ordinary phenomena of the weather may be recorded with certainty and brevity." The logs quantified the sea, taking what shipmasters and naval officers had always done, and formalized it in a systematic and codified way. The abstract logs brought order to the method and process of scientific observation in the ocean environment and, in doing so, began to transform the sea itself into an ordered place that could be understood and then represented. Maury had constructed a cyclical system of scientific research in which the observatory vetted the data submitted by mariners and returned it to them in continually revised editions of *Wind and Current Charts* and *Explanations and Sailing Directions* at no charge provided they continued to submit logs. Such a system was unprecedented in scope, if not in kind. It marked a unique relationship between science and the maritime community mediated by the navy that transformed the antebellum maritime world.[18]

The abstract log, then, required a sustained and systematic engagement with the marine environment on the part of Maury's mariner-observers. To page through the many thousands of these logs is to see an imaginative transformation in process. By the 1850s, Maury bragged that he had at any one time several thousand ships taking observations for him in waters all over the world. They came back to the observatory, in many cases, filled into the margins with notes that were vertical, horizontal, and on every plain between. In addition to the required empirical data, mariners often felt moved to write notes to Maury expressing their admiration and gratitude. "Capt. Chase desires me to say to Lieut Maury that the new route meets

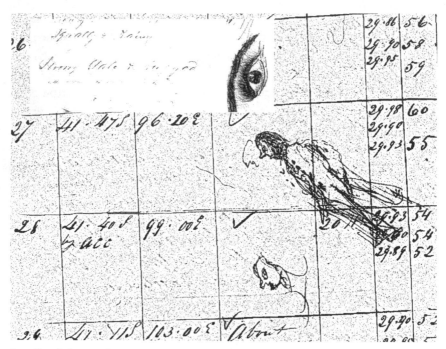

Details of abstract logs sent to Maury at the Naval Observatory. Eyes are prominent amid environmental data collected by shipmasters. Courtesy of the National Archives and Records Administration.

his approval in every respect and thinks that his passages have been shortened many days by the use he has made of the charts," a passenger aboard the brig *Georgiana* wrote Maury in 1851. "On examining your Sailing Directions and charts," wrote William Brewster, master of the ship *Contest*, "I can say I consider them the best guides ever given to the Navigator in pointing out the means of shortening the passage from New York to [San Francisco] and avoiding calms which causes so much detention." Many offered apologies to Maury for their failure to meet his standards of completeness for various reasons. Some logs came to Maury with doodling on them. More than one log depicts what appears to be a shipmaster whose most conspicuous feature is his eyes, perhaps a reflection of a favorite and meaningful phrase of Maury's, "the eye was successfully addressed . . . by a mere glance at the chart." If the chart's imagined power derived both from making visual the obscure and recasting the ocean in new ways in the mariner's mind, it seems mariners soon grasped the relationship between the many forms of cartographic vision and the transformations of knowledge coursing through the American maritime world.[19]

By the early 1850s, Maury's "Fair Way to Rio" had become one in an expanding series of new charts he titled the *Wind and Current Charts*, which together marked a revolutionary moment in maritime cartography for the new ways he represented the natural world on paper. Maury divided the charts into six series lettered "A" through "F." One series each documented ship tracks, trade winds, winds and currents, water temperature, meteorology, and whales. Multiple charts within each series covered the Atlantic, Pacific, and Indian Oceans, so that by 1860, the boundless seas of the world were boxed, lined, quantified, and filled with symbols, each calculated to express certain laws of nature as Maury surmised them from the many logs that now flooded the observatory.

The charts were, first and foremost, practical, but they also evinced new cartographic conventions that made the *Wind and Current Charts* different from the hydrographic charts produced by the Ex. Ex., set as they were in the traditional framework of marine navigation. Maury's work was based in the understanding that the winds and currents of the sea could "wreck or save the mariner" and "hasten, or delay him on his voyage according to his knowledge of them." By 1855, the Navy Department estimated that the *Wind and Current Charts* saved the American maritime sector "several millions a year" in shorter voyages. Mariners hailed Maury's charts as "one of the most valuable inventions of the age" and "the best guides ever given to the navigator." The charts offered new cartographic perspectives, taking the "pictorial conventions" that characterized the Western cartographic tradition and that typified land mapping and hydrographic charting, and adding to these new ways of representing natural forces that were often invisible, or nearly invisible, to the human eye. The *Wind and Current Charts* reconceptualized the sea, breaking it down into new structures, representations, and meanings.[20]

In the track charts of Series A, Maury set down the passages of all the ships for which he had data, exposing the triumphs and follies of individual navigators and infusing the sea with a history of American and—increasingly—international voyaging. For these charts, he sought "1000 tracks for every ocean." Based on this mass of information, he marked the average route between ports so that the mariner could identify the one by which he would have the best chance of favorable winds and currents. Once plotted together, Maury remarked, American ships seemed to be "cutting up the ocean in all directions." The busy routes of maritime commerce emerged from blank cartographic space, suggesting that far from a "waste of waters" or a "trackless waste" (the phrases that appear most commonly

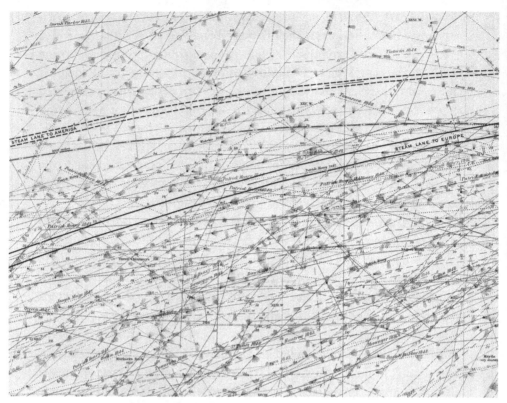

Detail of Maury's *Wind and Current Charts*, Series A, 1849, showing ship tracks, wind speeds and directions, and Maury's steamer lanes to Europe and America across the North Atlantic. Charts like these infused the open ocean with narratives and histories, as Maury said, for the benefit of all. Courtesy of the Library of Congress, Geography and Map Division.

in mariners' observations about the open ocean), the sea could be inscribed with many meanings, human and natural. At sea, the ship's foamy wake soon disappeared, but on the track chart, it remained a testament to the environment and a record of the mariner's encounter with the sea.[21]

The track charts showed that some vessels made speedy voyages while others plodded or wandered, but Maury deemed all revealing. "I find that tracks of vessels at sea are full of meaning," he wrote. "We have got so that we judge by them the character of Captains," he continued, "a crazy fellow always makes a crooked track." It was a simple axiom and one that was easily graspable by mariners themselves as they scanned the charts to and from their destinations. Maury preferred ordered linearity to the circuitous voyages of experience or folly. "Who were these engineers that laid out" such irrational "highways upon the sea," Maury wondered. In some seas, he had plotted so many tracks that the charts became a crowded mass of lines, suggesting, in an abstract way, that the mariner was not alone even as he saw no sail on the horizon. Indeed, Maury sounded this very point in his *Explanations and Sailing Directions*. "The object," he stated, "is to give every Navigator the benefit of the experience of all." With the track chart spread before him, he would know the conditions "his predecessors may have encountered in the same region and at the same season of the year."[22]

The pilot charts of Series C were part practical, part spectacle, constructing a sea whose winds the mariner could forecast with much greater certainty based on the past experience of so many mariner-observers. Maury divided the sea into grids, which he filled with numbers according to the frequency with which winds had been recorded on each point of the compass and in every month of the year. Maury reworked the common system of latitude and longitude, creating smaller squares of five degrees in which he quantified the winds. They were stunning, and perhaps even overwhelming, when viewed as a whole. But when taken grid by grid, Maury's system nevertheless conveyed an extraordinary amount of information in an immediate and straightforward way. Maury "aimed to get at least, on the average, 100 observations for every month in every district," or twelve hundred records for every square on the chart. He did not always achieve this; some seas were busier than others. But the pilot charts represented a significant leap in mariners' understanding of ocean winds.

The pilot charts had thus turned navigation from chance, hard-learned experience, or reactionary decision making into calculations of probability. For the first time, Maury bragged, the navigator "may examine his chart, and with such probability tell how the winds are, or at a given time will be in any part of the wide ocean . . . he may bet upon the prediction, and state in definite numbers, the chances for [or] against him." While the track charts rooted the sea in the past—delineating so many historical voyages of American vessels over the world's oceans—the pilot charts drew on that same past, but looked to the future, so that the mariner might know what his vessel would likely encounter when the sun rose again over what had long seemed an interminable expanse of waters. In short, these pilot charts turned the practice of marine navigation from one reactive to the forces of nature toward a proactive harnessing of winds and currents that would not impede voyaging, but aid and hasten it. To Maury, the sea was an ally designed by nature—and God, which I discuss below—to be used by those who went down to the sea in ships and did business on great waters.[23]

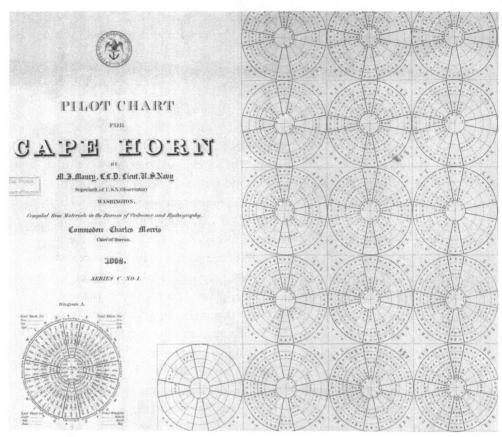

Detail of *Pilot Chart for Cape Horn*, 1852, showing the frequency with which winds had been reported from different directions in each month of the year. Courtesy of the Library of Congress, Geography and Map Division.

In the whale chart of Series F, Maury tracked the migrations of sperm and right whales, indicating the richness of certain cruising grounds at particular seasons even as whale populations declined. In the whale chart, Maury declared, "we shall be enabled to show whalers exactly when to go and where to go, to get into the midst of [whales] at any time of year." For the American whale fishery, this seemed a remarkable proposition. Maury indicated the animals' prevalence by using breached whales as symbols. The spout, as any whaler easily grasped, identified the type of whale—two spouts for the right whale, one falling forward for the sperm. Maury charted no other species than these two favorites of the American whale fleet. The value of the chart lay primarily in its commercial potential for these whalers as they pursued leviathans northward into the Bering Sea and thus into more extreme marine environments. American whaling was a business predicated on volume. Whaleships did not usually return until their holds were filled with barrels of whale oil or vast groves of baleen. The whale chart thus promised not only to show mariners where to go to find whales, but also held the prospect of changing the nature and duration of whaling voyages.[24]

To the profit-seeking whaling captain whose own livelihood derived from a percentage of the voyage—called a lay—the whale chart was also a visual spectacle. In the journals of whalers, the logbooks of whaleships, and—it should be said—also in the pages of Maury's abstract logs, whale symbology abounded, marking the moments when whales were seen, chased, taken, and tried-out—the term for rendering whale fat into oil aboard ship. In many cases, whalers inscribed these iconic bodies of whales with the numbers of barreled oil they rendered, thus commodifying nature in the profits of maritime capitalism. Maury incorporated this familiar iconography into his cartography. On the chart, whales seemed to stretch from ocean to ocean and coast to coast everywhere, breaching and beckoning the whaleman to lower his boats. They seemed ripe for the taking. While whaling did effect discernible changes in ocean biology, the primary environmental change here was imaginative. Just as settlers improved the land and planted it in wheat, cotton, and other agricultural products, the sea too emerged in Maury's whale chart as a watery field for resource extraction. This, of course, was more image than reality. Whalers knew they could not expect so many breached whales lying in wait for their harpoons. Maury labeled each whale with small letters indicating the season in which it would likely be found, so that a whale in any particular sea did not indicate its presence in perpetuity. But this took little away from the sense that the sea seemed full of whales such that "whale men," Maury quipped, "might well afford to give us a perpetual lay in all their ships."[25]

It is no coincidence that the decade of the 1850s witnessed Maury's whale chart, the height of American whaling, and the great whaling novel—arguably one of the greatest works of literature penned by an American—Herman Melville's *Moby-Dick*, published in 1851. A book about one whaling captain's vengeful quest "to seek out one solitary creature" seemed like a fantastic proposition. But as Melville scholars have shown, the pursuit of Moby Dick was grounded in many stories—Jeremiah N. Reynolds's own *Mocha Dick* among them—at the liminal spaces between fiction and reality, which watery settings seemed to so often convey. In chapter 44, "The Chart," Ahab pores over some wrinkled charts in his cabin, the folds of the

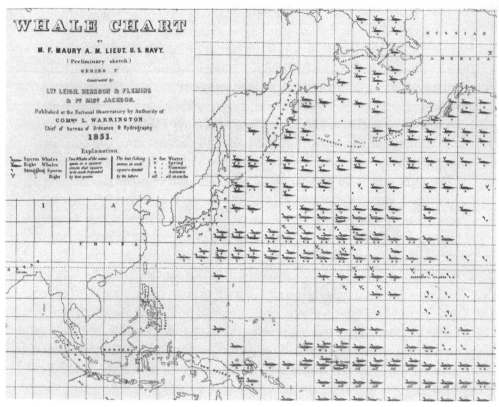

Detail of *Whale Chart*, 1851, showing sperm and right whale sightings in the western Pacific Ocean by season and location. Courtesy of the Library of Congress, Geography and Map Division.

paper reflecting Ahab's own wrinkled brow and thus the connection between the chart, the physical environment, and the imagined one as Ahab descends deeper into madness. Melville tells us that Ahab "knew all the sets of tides and currents" through "reasonable surmises almost approaching to certainties." An asterisk takes the reader to the bottom of the page, where Melville acknowledges that just such a system of research was underway, the work of Lieutenant Maury at the Naval Observatory. Just as Maury's charts told stories about the voyages of vessels, the courses of winds and currents, and the seeing and taking of whales, so too did Melville's narrative invoke naval science and cartography to lend credence to the notion—which Melville perhaps also sought to critique—that the ocean could be understood and that Ahab might one day find his whale "in the unhooped oceans of this planet."[26]

Maury's *Wind and Current Charts* rendered the natural world a work of design that obeyed universal laws and principles for ocean scientists to discern. In Maury's hydrography, the chaos of wilderness gave way to immutable laws of nature. The charts, he declared, dispelled "all doubt and perplexity." The sea "is never once left to the guidance of chance," he said, but rather the ocean and the atmosphere above were "obedient to law, and subject to order in all their movements. Though unstable and capricious to us they may seem," he continued, they operated "with regularity, and perform their offices with certainty." Maury saw order in the natural world, an order that—though he also attempted to convey it in words—could best be represented in cartographic form. It was in the chart that Maury could affect a change in the minds of mariners—not only to make them better navigators, but also systematic observers of the natural world and participants in a scientific project that would transform the ways Americans thought about the ocean and their place on it.[27]

Of course, Maury's system and his charts drew on the observations of and benefited from an international maritime world. Maury quite consciously framed the ocean as a "common highway upon which each society, like every nation, may make its ventures, and return in vessels laden with treasures to enrich the mind and benefit the human race." He worked closely with Sir Francis Beaufort, hydrographer of the Admiralty, whose own system of wind classification—the Beaufort Scale—was rooted in the same scientific ethos of empiricism and rationalization of nature as Maury's. Beaufort worked with Maury to arrange the first International Meteorological Conference in Brussels in 1853, which not only established international cooperation, methods, and standards based on Maury's system but also approved the use of Beaufort's scale. How, then, could Maury's work be so elemental to American maritime emplie if it was at the same time so international and founded in cooperation among so many seafaring nations?[28]

While these ocean highways and racetracks were held in common and Maury solicited abstract logs from and distributed his charts to an increasingly international maritime world, the effect of the observatory's work was to stimulate the expansion of American commerce over the world's oceans. "I have been with [American masters] in all parts of the world," Maury wrote in 1849, referring to his study of their logs, "and . . . I have found the American sea captain above the ship masters of all other countries is known for his enterprise and intelligence. America excels the world not more in her merchant ships than in her merchant captains," he told the maritime communities of New Bedford and New London, and he conjured in his mind, if not also on his charts, "a fleet of ships manned by thousands of American seamen all acting harmoniously together." Maury's common highway could be distinctly nationalistic. American ships, shipowners, and shipmasters whose commercial fleets before the Civil War were second only to Great Britain's would seize the commercial reins of a maritime empire—all navigational knowledge being equal. For Maury, those reins harnessed the winds and currents of the natural environment itself.[29]

Maury's charts everywhere effected changes in the American maritime world. He advocated new routes across the North Atlantic according to "Great Circle Sailing" that followed the spherical curve of the Earth rather than the linear compass headings that had long linked point of destination from point of departure. By 1855, he transcribed steamer lanes across the North Atlantic on the Liverpool to Sandy Hook passage—paths he called "the English" and "the American road-way." American clippers raced one another around Cape Horn to San Francisco, bearing gold seekers, entrepreneurs, and the spirit of Manifest Destiny. When the clipper *John Gilpin* made the passage in ninety-three days using Maury's *Wind and Current Charts* in the fall of 1852, it cut nearly in half a voyage, which had averaged 187.5.[30]

Reducing the duration of voyages in this way saved time, money, and lives, and won Maury the loyalty and respect of a maritime community that had initially been suspicious or indifferent, but it was Maury's active courtship of American commercial and naval captains and his estimation of them as scientific observers with the capacity to play an integral role in the process of scientific inquiry that most exemplified the changes his cartography brought to nineteenth-century seafaring. Mariners, Maury admitted, were "a much abused class," not least in the eyes of the scientific community. Sailors typically were perceived as the dregs of Euro-American society and their masters and mates, though a class above, remained outsiders oriented to the sea and not the intellectual centers of the nation's colleges and learned societies. Maury's system seemed to betray the professional impulse then coursing through American science, and, at the very least, it upset proper notions of class in early Victorian science. Scientists and sailors did not commingle. But Maury, who was both a naval officer and a self-taught scientist, viewed such relationships through his own lens. In his private and published writing, he began to refer to mariners as a "corps of observers" and his "co-laborers" in science. This was quite a turn from his previous rebukes. "I hold every properly qualified navigator to be a philosopher," he declared unequivocally in his *Explanations and Sailing Directions*. In part, this about-face was an affectation. In his private correspondence, he could be utterly dismissive of "the ordinary run of seafaring people." But Maury was a scientific outsider himself and, as such, shared some affinity with the sailor. The *Wind and Current Charts*, in Maury's estimation, exhibited the "intelligence and public spirit" of the American mariner. "As a sailor," he added, "I mention it with proud satisfaction." Maury understood that no other American knew the sea as well as the mariner. Whether Pacific whaleman or captain of a Liverpool packet, life and prosperity at sea demanded a certain store of environmental knowledge that made the mariner a natural observer of the environment.[31]

There is evidence to suggest that Maury's abstract log and his *Wind and Current Charts* made mariners more keen, systematic, and scientific observers and users of the natural world. "I have determined, during the coming voyage, to keep the 'Abstract Log' of Lt. Maury," wrote John Young, master of the merchant ship *Venice*. By doing so, he intended to "add my mite to the cause of science" that "navigation shall be so simplified, and reduced to 'fixed principles,'" and "that all uncertainty may be removed." Though mariners already knew the sea well, the abstract log encouraged them to see it in a different way. Captain Foster of the ship *Creole* bound from Le Havre to New Orleans in 1849 became fascinated by "a species of Mollusca." With the "aid of a microscope," Foster proceeded to observe, examine, and dissect the creature, all the while describing to Maury its physical features. Captain Phinney of the ship *Gertrude* wrote to Maury in 1855, praising the way "your great and glorious task" had changed his relationship to the sea. This, Phinney explained, was the task "of teaching us sailors to look about us. . . . For myself, I am free to confess that for many years I commanded a ship, and, although never insensible to the beauties of nature upon the sea or land, I yet feel that, until I took up your work, I had been traversing the ocean blindfolded. I feel that . . . you have done me good as a man. You have taught me to look above, around, and beneath me, and recognize God's hand in every element by which I am surrounded." Aside from the explicit references to natural theology, these accounts bear testament to the ways Maury's system could change the nature of mariners' encounters with the marine environment.[32]

Yet however transformative Maury's cartography proved to be both on deck and in the minds of mariners, Maury, like Wilkes and his surveyors, still struggled to portray the complex movements of a dynamic environment on paper. Unlike Wilkes, though, Maury had diverged from the traditional conventions of the hydrographic chart in hopes of communicating and imparting new meanings, but he found both the chart and the mariners who read them frustrating tools of his cartographic authority. The *Wind and Current Charts* amassed such a storehouse of data about the marine environment that it often appeared on the charts themselves as an almost unintelligible mass of knowledge that seemed to undermine Maury's primary belief that his charts were powerful visual tools intended to be "a fine show" and to "strike the eye at once." Of course, such a multitude of tracks might have sacrificed clarity, but Maury went to pains with his engravers and lithographers to produce charts in a system of colors and solid, dashed, and dotted lines to differentiate seasons and months. "I *must* have the 4 colours, and you must give them to me without offending the eye too much," Maury demanded in a letter to one of his lithographers. "Books," he wrote to former president John Quincy Adams, "impart information through the ear—these charts through the eye." Maury's demands on lithographers, on antebellum print technology, and on the inherent limitations of the cartographic medium to convey the sorts of meanings he intended were daunting challenges that he only partly overcame. As Maury grudgingly realized each time his lithographers could not meet his demands, the chart itself had limitations whose boundaries in representing the natural world with fidelity—in all its complexity and movement—could only be pushed so far.[33]

Maury also grappled with problems of interpretation. As his observation to Quincy Adams makes clear, Maury appealed to the mariner's sense of the visual. He intended his charts to be spectacles. Yet, as Maury learned, the mariner's eye was a subjective lens that did not always see the ocean as he intended. In addition to the passages of individual vessels, Maury's track charts had designated the mean passage from port to port to show the navigator the average course of all vessels. Navigators, Maury remarked, "have inferred . . . that [these lines] must be followed as rigidly and as closely as though they marked out a channel-way, on either side of which if a vessel should fall, she would find herself in difficulty." Indeed, by the 1850s, Maury had marked just such channel ways for steamer lanes across the North Atlantic. But to Maury's dismay, some navigators had trusted so wholly in his work, that they followed these lines unequivocally. When contrary winds and currents sprang up unexpectedly, some mariners stuck to this route, beholden to the idea, as Maury put it, that "there is some sort of virtue in the black mark on the chart." Maury had charted himself into a contradiction. On the one hand, his cartography and his texts had worked to transform mariners' ideas about the sea and many embraced its order, its laws, and its design. But he knew that the sea did not always follow his rules. "I do not claim for vessels on the new route an exemption either from head winds, baffling airs, or calms," Maury admitted. "On the contrary, I expressly show that vessels on the new route are liable to all these. Nor do I claim for the new route short passages *invariably*. I only claim that the average of the passages by the new route will be shorter than the average of the passages by the old." Maury had thus achieved his expectations, but at the cost of the misapprehension that the sea always worked in certain ways when, in fact, it did not.[34]

Nevertheless, command of the observatory and the publication and popularity of the *Wind and Current Charts* had elevated Maury to the highest levels of the American scientific world from which he espoused ideas about the divine order of nature that sat uneasily with a growing chorus of opponents in the American scientific community. The origins of this rivalry, particularly between Maury and Joseph Henry, secretary of the Smithsonian Institution, and Alexander Dallas Bache, who succeeded Hassler as superintendent of the Coast Survey, are complex and multifaceted. As many historians of science have proposed, there was certainly professional envy on the part of Henry and Bache for a naval officer with no formal education in the sciences, commanding a world-class observatory in the nation's capital, where he enjoyed proximity to congressional purse strings, the attention of the American people generally, and, of course, the increasing support of a maritime community that before the Civil War was one of the primary drivers of the American economy and generators of revenue for the federal government. There was also a spat with Henry over the intellectual discoveries of an astronomer whose work at the Observatory had been published under the Smithsonian's name when he took a new position there. Finally, Bache, whose great-grandfather was Benjamin Franklin, had clashed with Maury over hydrographic jurisdictions, particularly which organization had the purview to study the Gulf Stream that the elder Franklin had done so much to outline.[35]

Aside from political and hydrographic turf battles, Maury practiced a brand of ocean science deeply immersed in the mechanical efficiency of the mid-nineteenth century and, at the same time, rooted in the Bible, which, in its appeal to popular audiences, seemed to Bache and Henry to transgress the progress they deemed so important to the professionalization of American science. To Bache, Henry, and others, Maury's wordy pronouncements about "the developments of order and the evidences of design" seemed to pull science back into the realms of the amateur naturalist whose knowledge might have been broad, but whose expertise in any one of several emerging fields was shallow at a time when these men were attempting to move American science to more firm professional ground.

Influenced both by the Industrial Revolution and his religious faith, Maury conceived of the ocean as an interconnected divine mechanism, which infused his prose and ideas with a lyrical eloquence unmatched by many scientists of the era, but which also existed uneasily with the dispassionate secularism of an emerging scientific profession. A favorite analogy of Maury's was to compare the order of nature to the order and efficiency of a machine. "The atmosphere," he wrote, was "a vast machine, that is tasked to its utmost, but . . . one that is always in order and never breaks down." Again—"what a powerful machine is the atmosphere . . . as obedient to law as the steam engine to its builder." Pressing the metaphor further, Maury likened the Gulf Stream to a natural furnace system. The tropics, he explained, were the furnace itself, heating water in the "cauldron" of the Caribbean Sea and the Gulf of Mexico. The Gulf Stream, in this analogy, was "the conducting pipe," conveying warmed water into the "hot-air chamber" of the North Atlantic. As historian D. Graham Burnett has suggested, there is a parallel between this analogy and the new state-of-the-art furnace system that Maury had installed in the observatory. He was quite clearly and personally informed by notions of progress and efficiency made possible by industrialization in America. Maury's natural machinery was benign. He thought that it evinced the grand wisdom of the natural order just as the machine had symbolized, to some Americans, the triumph of human ingenuity. In Maury's mind, the machine was not just in the garden, it was the garden itself.[36]

This "exquisite machinery" pointed to the providence of God, who, in Maury's natural theology, created the winds and currents along with the Earth, the sea, whales, and all other creatures over which humans were supposed to exercise control. A devout Methodist, Maury saw no contradictions between science and religion. "The right-minded mariner," he counseled, "hears His voice in every wave of the sea . . . and feels His presence in every breeze that blows." He cited Job on gravity and Solomon on atmospheric circulation, finding in these biblical verses general truths that affirmed his own research. He firmly believed that science could, in fact, lead him closer to God. "As our knowledge of Nature and her laws has increased," Maury argued, "so has our understanding of many passages in the Bible been improved." To Bache, Henry, and others, Maury's natural theology seemed in contradiction with the Enlightenment rationalism of science.

One can almost sense their exasperation when Maury launched into another of his biblical metaphors. Indeed, his reputation at the hands of more recent historians of science has suffered, not least, because of his reconciliation of religion and science. Nevertheless, Burnett and other historians of nineteenth-century science have pointed to the "durability, complexity, and significance of natural theological thought in the history of the sciences of matter, motion, and life." Maury and other eminent ocean scientists of that era such as Charles Darwin and James Dwight Dana had few qualms about the convergence of God and science. At midcentury, the two were far from mutually exclusive.[37]

By the mid-1850s, the rift between Maury on one hand and Bache and Henry on the other—between naval and civilian professional science—had opened wounds that spoke not only to institutional and scientific jealousies but to the broader question of who could claim to be a scientist within an increasingly circumscribed profession. In question, Burnett has put it, was Maury's aim "to marry the knowledge of nature and the knowledge of the masses." By 1857, Maury had attempted to apply his system of scientific observation at sea to the continental United States, hoping to enlist farmers to the cause of meteorological observation that would benefit both American agriculture and science. Here, Maury incurred the ire of Henry, who was leading a similar Smithsonian program of meteorological study that did not rely on Maury's crowd-sourced data gatherers, but on a more select corps of observers working with Henry and coordinated by the Smithsonian. An anonymous editorial in the February 18, 1857, issue of the *Boston Atlas*, which Maury attributed to Henry's influence, railed against Maury's supposed amateurism. "Even half-educated people," the anonymous writer rebuked, "should protest against our being held nationally responsible for the character of the essays which are ceaselessly issuing from the 'Hydrographical Office.'" True scientific research, he suggested, relied on "systematic records" from "carefully compared instruments" taken by "learned men" in the nation's "seminaries of learning. I had always supposed," the writer concluded, "that educated men were more likely than ignorant ones to deduce correct results even from data of equal value." Maury was a scientific democrat, eschewing the pretensions of Bache and Henry in favor of science carried out by people deemed outside the professionalizing scientific community. At root, the rivalry between Maury and Bache and Henry emerged from conflicting and irreconcilable definitions of science in nineteenth-century America. To Maury, Bache and Henry's science smacked of elitism and exclusivity. To Bache and Henry, Maury was a popularizer whose theories sometimes did not hold up to scrutiny and, in some cases, diverged problematically from a growing scientific consensus.[38]

In the 1850s, Maury also moved from the practical science of navigation and hydrography to larger theoretical arguments concerning the movements of winds and currents that garnered him the title "the father of oceanography," but also provoked the suspicion of much of the American scientific community. In 1855, Maury published his most well-known work, *The Physical Geography of the Sea*, whose title he borrowed from Humboldt and in which he synthesized the larger scientific results that he had gleaned from almost a decade's time in the abstract logs of his mariner-observers. In many ways, it was an unprecedented text, setting the marine environment within the common interest of various scientific fields, and thus prefacing the emergence of the field of oceanography in the period after the Civil War. Yet the text also betrayed Maury's weaknesses as a theoretical scientist. In it, for example, he ascribed the circulation of the atmosphere to magnetism. This was a theory, as geographer John Leighly writes, "which did not convince even his lay critics." Even Maury's work on the Gulf Stream, which had done so much to advance knowledge of that current, exposed analytical flaws. Maury showed little regard for competing theories of the current's origin. Rather, he proposed that differences in temperature and salinity, rather than wind as others surmised, combined to propel warm water northward. Maury's conflicted legacy as a scientist is often pinned on the weaknesses of *The Physical Geography of the Sea*. Such indictments, however, magnify these problems at the expense of Maury's hydrographic and cartographic work as a whole. While he should rightly be held accountable for theories that ignored or dismissed evidence to the contrary in espousing his own theories, Maury nevertheless proved among the most important scientists of the antebellum era not because of the soundness of his ideas, but rather because of the practical and imaginative power of his charts and his ability to negotiate increasingly exclusive boundaries between the American maritime world, the scientific community, and the navy, all with a stake in the work of marine science and the extension of the American maritime empire.[39]

Like the voyage of the Ex. Ex., the navy, through Maury's work at the observatory, had enlisted science in the service of American nationalism and empire even as he appealed to international scientific cooperation in the study of the ocean. Literary allusions to wilderness frontiers aside, both Maury and his fellow naval officers framed science and hydrography more specifically within the expanding realm of American maritime commerce that was part and parcel of American antebellum expansionism. In the pages of the *Southern Literary Messenger* in July 1841, an anonymous fellow officer and supporter of Maury's hydrographic program praised his "active genius and Patriot's heart . . . to erect the beacon of hope amidst the darkest pavilions of the waters." He wrote that Maury's work did much "to alleviate the sufferings of the shipwrecked mariner," where "the winds and the waves" were "the only obstacles to the ceaseless floating of the star spangled banner upon every sea." By the 1850s, Maury himself was writing, "wherever commerce goes, it civilizes and with civilization wants are multiplied, and the ability to supply those wants increased." Maury's common highway was thus an imperial one, bringing trade, profits, and American civilization with it.[40]

While Maury was busy combing abstract logs, he also supervised from his post at the observatory several naval exploring expeditions that—like the Ex. Ex. before them—sought to expand the realm of commerce and science. There was an expedition to the Dead Sea in 1847 that, in part, sought to discover the biblical roots of Christianity, which reflected Maury's own larger sense of the divine order of nature. The North Pacific Exploring Expedition, 1853–56, was cause and consequence of the growth of American commerce toward Japan and the North Pacific as the American whale fishery pursued its prey into icy Arctic waters. Its mixed record of achievement in hydrography and diplomacy was obscured by the mental instability of its first commander—Cadwalader Ringgold, who had led Wilkes's assault on Malolo in 1840—and by the Great Chicago Fire of 1871, which destroyed much of its marine invertebrates collection, and, not least, by the coming of the American Civil War. By the early 1850s, Maury's hydrographic and scientific interests had increasingly become intertwined with the expansion of the Cotton Kingdom and his growing fears about the dissolution of the union he served. A southerner born in Virginia and raised in Tennessee, Maury oversaw an expedition to the Amazon led by his brother-in-law, Lieutenant William Lewis Herndon, which spurred him to advocate American expansion into South America based on a climate favorable to the growth of cotton and passages by sea between Brazil and the United States aided by favorable winds and currents. "The free navigation of that River," Maury wrote in June 1850 as Congress debated the Compromise of 1850, "is my remedy for preserving the Union." Maury thus employed an oceanic and climatological determinism in advocating an American empire within the context of similar ventures of Southern expansionism in the 1850s. His views represented a sort of environmental corollary and sectional counterpoint to the arguments of Unionists such as Senator Stephen A. Douglas who had long hoped that climate would hem the spread of slavery where American politics apparently could not.[41]

Another effect of Maury's cartography was to make the open sea and ocean depths a meaningful place for scientific inquiry and commercial activity. As Helen Rozwadowski has argued, the mid-nineteenth century was a time in which scientists discovered the deep sea, and Maury was a central figure in this moment. She adds the sea as destination for scientific inquiry to the theoretical binary of ocean as barrier and ocean as highway that has long informed the historiography of maritime history. Through Maury, the navy was an important actor in the discovery of the deep sea. In 1849, Maury had managed to secure the use of three naval vessels to "make observations upon the winds and currents of the sea and to collect other facts in connexion [sic] with the 'Wind and Current Charts.'" The first of these, the unseaworthy schooner Taney, had achieved what Maury thought to be a momentous accomplishment. On November 15, 1849, Taney's crew sounded to a depth of 5,700 fathoms, or 34,200 feet, deeper than any previous cast and, indeed, far deeper than Maury thought the ocean floor to be. But soundings by the second of these vessels, the brig Dolphin, proved the 1849 cast to be erroneous. Beginning with Dolphin's 1852 cruise, however, Maury employed a new deep sea sounding instrument developed by one of his officers at the observatory, which promised more accuracy and to bring back a sample of the sea floor for examination.[42]

With more success and greater precision, Dolphin's crew used the new sounding device as it crisscrossed the Atlantic in 1852 and 1853, revealing a new picture of the sea floor that figured significantly in Maury's ideas about a benign sea created for maritime enterprise. The new machine was the invention of Lieutenant John Mercer Brooke, whose revolutionary contribution to the history of sounding technology was the detachable weight. Brooke used thirty-two pound cannon shot, and sometimes heavier, to pull his sounding wire to the bottom, where it detached from the weight and, in a tallow-filled cylinder, brought up sediment from the bottom for investigation. With the detachable weight, Brooke had solved the greatest quandary of deep sea sounding—how to know when the lead actually touched bottom. Previous leads, light enough to be heaved back aboard ship, were subject to the caprice of undercurrents, which carried them horizontally instead of at the intended right angle with the sea floor. With Brooke's device, Dolphin's crew sounded the Atlantic at intervals of two hundred miles.

These measurements gave Maury data for a bathymetric chart, which showed, for the first time, a vague outline of the Atlantic sea floor. The effect of the new technology was to transport Maury from the halls of the observatory to the deep sea floor itself. With Brooke's lead, Maury exclaimed, "I have been in the depths of the Ocean." To the mariner and the scientist, Maury wrote in his *Explanations and Sailing Directions* that the bottom was "quite as irregular in its outlines, in elevations and depressions, in its mountains and its valleys, as is the face of our continents." In the North Atlantic, however, Maury stumbled on a stretch of sea floor that, he thought, suggested something providential. At a depth of twelve thousand feet, a nearly flat bed of shells appeared to stretch from Newfoundland to Ireland. Maury shrewdly designated it the "Telegraph Plateau." His ongoing correspondence with Cyrus Field, proprietor of the trans-Atlantic telegraph cable, suggested the practical value of deep sea hydrographic surveying for commercial purposes, even as Brooke's deep sea device brought up deep sea ooze and other substances of special interest to science.[43]

Laying a telegraph cable across sixteen hundred miles of ocean was a daunting proposition at midcentury, requiring all the ingenuity, knowledge, and publicity available to finance and execute it. Maury was a master of all three. As a result of *Dolphin*'s cruise and other deep sea soundings by naval vessels using Brooke's device, Maury had written to Field in 1854 that the cable was practicable. But Maury's illustrations and his pen performed perhaps an even more important function. Rozwadowski contends that, once publicized by the newspapers and periodicals of the day, Maury's hydrography "presented an attractively benign picture of the depths." He wrote to Secretary of the Navy James C. Dobbin that the sea floor of the Telegraph Plateau was "quiet . . . as a millpond." The press drew on these words and cited samples from the bottom raised by Brooke's device to popularize the venture, declaring that the sea floor was "quiet and undisturbed," and "a sort of bed of down for the cable to rest upon." Here, according to Rozwadowski, was a new idea about the deep sea. Previously mysterious, dark, and unfathomable, Maury's work and Brooke's sounding device began to recast it as a benign environment that fit quite well within Maury's larger cartographic ideology and, in opening up the deep sea to scientific inquiry, lent a new dimension to the navy's hydrographic program.[44]

The need for a new sounder and the difficulties Field faced in various attempts to lay a working cable, however, suggest that this kind of scientific research was not as simple and as transparent as Maury's pen, or even his charts, suggested. The problems were technological, environmental, and human. Brooke's device solved some festering problems of deep sea sounding technology, but it was far from perfect. Among other things, it required an extended period of calm seas. The line itself, with heavy shot, was prone to part mid-cast as Brooke discovered firsthand on a surveying cruise in the North Pacific. One officer voiced the concern of many when he remarked that "deep sounds will, I think, always be attended with great uncertainty," and particularly "if there should be a current." Other methods of investigation also remained primitive. In 1843, Maury had advocated for a general study of ocean currents by suggesting that mariners throw bottles overboard with their position enclosed. When picked up, he hoped, mariners would return them to the observatory for analysis. Elsewhere, he suggested that subsurface currents of the Gulf Stream might be identified using a weighted canvas parachute suspended by fishing wire and corks. The sea itself, of course, was a dynamic environment whose processes Maury only partly understood despite the mass of data mariners had collected for him. His bathymetric chart of the Atlantic sea floor, which he constructed from these deep sea sounding voyages—for all its cartographic significance—nevertheless relied on a handful of soundings and presented only a very basic representation. Despite all the certainty that suffused Maury's charts and his own pronouncements, the deep sea remained mostly unknown.[45]

If Maury's scientific record raised questions about who could claim the title of professional scientist, the naval officer corps went through a similar crisis of identity in the 1850s that set Maury, among others, in its crosshairs. The Naval Efficiency Board of 1855 was created by an act of Congress and was charged with thinning the navy's ranks of officers not fit to achieve the highest duties of the naval service, that is, the command of a ship or squadron of ships at sea. The board also met at a time of promotional stagnation and amid questions about the continued relevance of a system of promotion based primarily on seniority and not merit. There were too many officers and not enough ships to command them. Little accordance was given to science, which remained a secondary expertise within the service despite the fact that the navy had emerged as one of the most important scientific institutions in antebellum America through various exploring expeditions and Maury's leadership in hydrography and astronomy at the observatory. To his disbelief, Maury was one of more than two hundred officers whose careers were affected by the board's decision. While not cashiered from the service, Maury was put on the reserve list and given half pay.

Maury believed that the board's verdict in his case, though ultimately overturned by Congress with the support of Maury's influential friends, smacked of the tenuous place of science within the navy's ranks. He sensed it was not so much about his bum leg, which deprived him of the ability to regularly go to sea between stints on land, but his work as a scientist that was not valued within the service. "I have, without cause, been made to suffer a grievous wrong," he opined. He condemned the board as "a monstrous inquisition" made up of officers "not one of whom has the least pretensions to any scientific attainments." Maury petitioned the board, the secretary of the navy, and Congress for redress, citing, in histrionic fashion, British admiral Horatio Nelson, who had fought the Battle of Trafalgar in 1805 with one eye and one arm. But here was the crux of the issue. Maury was not Nelson. He differed from most of his fellow officers in his preference for the halls of the observatory over the quarterdeck of a ship in battle. Ultimately, Maury's influence compelled his reinstatement in 1857 with a promotion to the rank of commander retroactive to 1855. Nonetheless, the affair suggests Maury's conflicted status within the navy's ranks. The board's findings had, for a time, identified his disability as more significant than his research.[46]

Ultimately, neither Maury's reinstatement on the navy's active list nor his schemes to expand the Cotton Kingdom into South America could save his career from the existential crisis in which many military officers from Southern states found themselves in 1861 when eleven states seceded from the Union to form the Confederate States of America and fought a four-year struggle over the meaning of liberty, federal governance, slavery, and freedom. Maury resigned his commission in the U.S. Navy—along with John Mercer Brooke—and, as one of the incipient Confederate Navy's most experienced and accomplished officers, turned his attention to developing underwater mines and contracting with British shipyards for fast commerce raiders waging *guerre de course*—or commerce warfare—against Yankee merchant shipping. Maury's career in the U.S. Navy was over and, with it, his ambitious program of hydrographic and scientific research that had made the American merchant and whaling fleet—ironically, the very same vessels he now hoped to destroy as an officer in the Confederate States Navy—second only to Great Britain in the antebellum era.

Nevertheless, he left behind a record in science, naval affairs, and maritime commerce that placed him among the most consequential figures in nineteenth-century America, transforming the deep, open ocean into an ordered commercial empire in which mariners seized a watery nature set down in the *Wind and Current Charts* to hasten their voyages. He was instrumental in constructing the sea as a watery racetrack and "a common highway," to borrow one of Maury's favored phrases. As his cartography enlisted mariners as observers and participants in the scientific process, he sought—in many cases with success—to dispel the experiential and folkloric meanings that had previously informed marine navigation in the United States by changing the ways that mariners visualized the marine environment both in material and imagined terms. His charts imparted new meanings on the sea. As they appeared on the charts, the tracks of vessels, the quantified frequency of winds, and the bodies of whales suggested to the mariner that far from a waste of waters or a boundless ocean, the sea held narratives, histories, and a larger, grander design whose exposition in hydrographic charts and texts expanded the scope of American commercial empire from the ocean's surface to the atmosphere above and the ocean floor far below. Maury's abstract log and his *Wind and Current Charts* fundamentally altered the way many mariners encountered, thought about, and used a natural world so integral to their livelihood, to the fate of their crews, their ships' owners and merchants, and indeed, an American economy rooted in maritime commerce more broadly. Even still, for all their unprecedented design, Maury found his charts sometimes unable to portray the complex dynamics of wind and water whose movements the world over sometimes overwhelmed the two-dimensioned chart's own capacities and that of the era's lithographic print technology. The chart, once again, could hardly pretend to cartographic comprehensiveness.

Nevertheless, Maury's work in hydrographic science won him an international reputation and sealed his place as a bridge between the navy, marine science, and the maritime world, even as he found himself on the margins of each of these communities at a time when professional concerns in all three diverged. Maury was a figure with one foot in each sphere, and so he was able to draw on all three at a time when American empire at sea was defined by the intersection of scientific, naval, and commercial imperatives. More than this, Maury's work at the Naval Observatory had fundamentally changed the way nineteenth-century seafaring Americans understood the ocean. "As one traveler in the wilderness follows in the trail of another," Maury reasoned, "so, it was discovered, did the trader on the high seas follow in the wake of those who had led the way."[47]

# 4 Conquering Old Ocean

· · · · · · · · · · · · · · · · · · · · · · · · · · · · · · · · · · · · · · · · ·

The haunts of the water gods are brought from the realms of
poetry into the straight lined chapters of fact and science.

—Unidentified American newspaper correspondent,
Yokohoma, Japan, May 15, 1874

In September 1873, a strange warship eased out of the Mare Island Navy
Yard in San Francisco Bay, headed north to the coast of Alaska, then across
the Pacific to Japan and back again, surveying a route for a trans-Pacific
submarine telegraph cable. USS *Tuscarora* was a wooden-hulled, three-
masted steam sloop commissioned in 1861. During the Civil War, it had
hunted Southern commerce raiders, the same ships the naval-scientist-
turned-Confederate-agent Matthew Fontaine Maury had contracted from
British shipbuilders. Now, though, no cannon protruded from *Tuscarora's*
gun ports save a few symbolic muzzles. The imposing eleven-inch Dahlgren
swivel guns that had once dominated its spar deck were gone, replaced with
the framed metal wheels and barrels of a Thomson Deep Sea Sounding Machine and miles of hemp rope and wire line. Over the guns' imbedded
tracks, the workers at Mare Island constructed a chart house. "A great
change was made in her appearance," observed the ship's writer Henry
Cummings. "The shot lockers," where ordnance had once been stored, he
noted, "were filled with sinkers of different kinds and sizes, the spar deck
was crowded with coils and reels of line and wire, and the chart house contained curious contrivances of different inventors for bringing up bottom
soil at great depths." *Tuscarora* was a curious-looking man-of-war, indeed,
neither warship nor scientific research vessel, but some "contrivance" of the
two.[1]

As *Tuscarora* left the Golden Gate astern and sailed into the vastness of
the Pacific, it entered an oceanic world being gradually transformed by science, technology, and new, but still nascent, notions of American power.
The prospect of laying a submarine telegraph cable between the United
States and East Asia—to "bring Yokohoma and Pekin within speaking distance of New York," as one newspaper put it—was a commercial project. So,

for example, was blasting a channel through the coral at Midway Atoll in the North Pacific for the Pacific Mail Steamship Company to recoal its steamers in the calm waters of Midway's lagoon. These joined many other pursuits from the Isthmus of Panama to the coast of Alaska and the Arctic in which commerce, science, and the navy were intertwined after the Civil War and whose antebellum precedents lay in Secretary Paulding's empire of commerce and science and Maury's common highway. Yet submarine cables and coral atolls also increasingly knit together new visions of American empire that were more muscular in tone, strategic in prescription, and territorial in definition. By century's end, this new empire stretched from the Caribbean to the Western Pacific and required a new, modern fleet, new strategic roles for science, and new militarized definitions of the natural world to defend it. When *Tuscarora* set out on its deep sea sounding voyage, however, all that lay in the future, but the roots of America's territorial empire were, in part, set here in the hydrographic work of the postwar navy. Alongside the diplomatic and naval precedents to the United States' turn-of-the-century imperialism long-established by Walter LaFeber, Kenneth Hagan, and others, we must acknowledge the work of naval surveyors and cartographers whose surveys, charts, texts, and rhetoric framed an era of important transformation rooted in changing ideas of and encounters with the ocean environment.[2]

From 1865 to 1890, naval scientists increasingly relied on technology to affect the scientific, commercial, and environmental transformations that reflected the gradual emergence of this new empire. The steam engine's increasing ubiquity in the postwar maritime world opened up new passages for navigation and generally rendered vessels less dependent on winds and currents, but more dependent on coaling stations. Steam engines also formed a vital component of new deep sea sounding machines such as the ones aboard *Tuscarora*, which Cummings noted with such detail and interest. Indeed, Cummings seemed to linger on the technical aspects of *Tuscarora*'s own metamorphosis from warship to deep sea surveyor, which suggests something of the centrality of new technologies not only to the operations of steam-powered vessels but also to the nature of hydrographic work and scientific inquiry at sea more broadly in this era. For Cummings and others, machines were spectacles whose power evinced human conquest over nature in remote and previously inaccessible waters. As the American empire expanded to more extreme environments that at times tested the limits of human endurance, technologies pierced great depths and remade marine environments. "Old Ocean, who for ages has stubbornly resisted all attempts to penetrate the secrets buried in the bosom of his waters, has been conquered," Cummings declared at the conclusion of *Tuscarora*'s transoceanic voyage, "and the way is now opened by which his innermost recesses may be explored." Such pronouncements spoke to the continued belief that the marine environment might be tamed and ordered by science aided now by technologies that seemed to render the natural world even more benign and beholden to American claims than Maury's antebellum highway. But if such rhetoric was commonplace in the postwar era, it also proved premature, speaking more to American aspirations—scientific, naval, and commercial—than to the reality of navigation and scientific study in a marine environment that not only retained many of its secrets but flouted pretensions to understand, represent, and alter it.[3]

The American Civil War marked a watershed moment for the U.S. Navy and the American maritime world. Despite a historiographical focus on land warfare, Union naval superiority off the Confederate coasts and its dominance of the inland rivers proved decisive factors in the war's outcome. Still, these so-called brown water operations, in which the navy navigated shallow waters, proved challenging even in ocean and riverine environments relatively well-charted by the U.S. Coast Survey, which had cartographic jurisdiction over American territorial waters and the Army Corps of Topographical Engineers, which charted American rivers. The U.S. Navy was no stranger to littoral operations. In the recent war with Mexico, 1846–48, for example, American vessels blockaded coastlines and ascended rivers to assault port towns that in many ways foreshadowed similar work during the Civil War, although American coasts and rivers remained better charted than Mexico's. The navy's combat operations in these environments, in important ways, relied on knowledge of the natural world set down on navigational charts. Since charting American coastal and riverine environments was not primarily the purview of the navy and since naval operations during the Civil War were generally confined to American territorial waters and thus did not speak primarily to American overseas expansion, I have left the topic of naval operations in the natural environment during the Civil War to future study. Naval war in the marine environment is a topic I take up more fully in chapter 5 in examining the navy's blockade operations during the Spanish-American-Philippine War at century's end.[4]

On the high seas, Confederate *guerre de course*, or commerce warfare, prosecuted in part by using Maury's *Wind and Current Charts*, devastated the Yankee merchant and whaling fleets. The most effective raider, Captain Raphael Semmes's C.S.S. *Alabama*, took sixty-eight American-flagged commercial vessels in a twenty-two-month voyage between 1862 and 1864. In his account of the war, Semmes praised Maury as the "chief blazer" for his scientific and commercial accomplishments, noting that "the most unscientific and practical navigator may, by the aid of these charts, find the road he is in quest of." For Semmes, the commerce hunter, however, Maury's charted roads pointed in the direction of Yankee shipping lanes. Semmes, in effect, turned Maury's charts against the maritime community that had benefited so much from them before the war. In February 1863, he brought *Alabama* to "the charmed 'crossing,' leading to Brazil," which Maury had identified on his first wind and current chart in 1848. When Semmes continued to the crossing of the equator "recommended to the mariner as being appropriate to his purpose," he found "a great many ships passing, both ways, on this road," but, he noted with disappointment, there were very few American-flagged vessels for the taking. They had left the "common highway," which Maury had "blazed" with his charts through the ocean wilderness, becoming "skulkers" and "rogues" in Semmes's mind. In fleeing these rationalized routes, American ships seemed to Semmes to have left civilization behind them and had themselves become uncivilized in the process.[5]

After 1865, the U.S. Navy returned to its historical role protecting and promoting American seafaring commerce, but this link, which had been so strong since American independence, became tenuous. The fleet decreased in size from a height of some seven hundred vessels in 1865 to less than two hundred three years later. During the war, Confederate commerce-raiders such as the *Alabama* had ravaged Yankee commerce in the Atlantic and the Pacific, raising insurance and freight rates and causing such a general fear among merchants and shipowners that many sought protection under other flags of registry. This "flight from the flag" became permanent after the war when the U.S. Congress barred such vessels from returning to American registry. The effect on the maritime world was dramatic. The percentage of waterborne cargo carried by American-registered vessels decreased steadily from 66 percent in 1861 to less than 10 percent in 1898, which, together with the increasing prevalence of steam-powered vessels less dependent on winds and currents, meant that Maury's common highway laid down on his *Wind and Current Charts* was neither so powerful nor so central economically and militarily.[6]

Yet even as American oceanic commerce shrank and the nation's attention turned toward pressing postwar questions of Reconstruction, the political and military will to support American commerce remained, now joined by emergent strategic interests. In 1867, Secretary of State William Henry Seward orchestrated the purchase of Alaska from Russia for $7.2 million. The same year, the steam sloop *Lackawanna*, under the command of the former Ex. Ex. officer William Reynolds, claimed the Midway Islands, approximately one thousand miles northwest of Hawaii. Seward had been an early advocate of American expansion into the Pacific, citing in 1852 "the subjugation of the monster of the seas to the uses of man." He meant whales and American whalers who were now pressed farther north into harsh, icy marine environments in search of their prey. "Our hardy whalesmen," lamented Congressman Thaddeus Stevens, "are obliged to double the Cape and make their years of abode in these inhospitable regions, where their own country does not own a foot of soil beyond the forty-ninth degree of latitude." Others in Congress called attention to Alaska's "soil, its harbors, its fisheries, its forests," urging their fellow representatives to remember "always that the natural tendency of the mind is to undervalue all unknown lands." Still others saw Alaska's acquisition as an opportunity to "cage the British lion on the Pacific coast. England's star has passed its zenith," declared Congressman William Mungen in 1868. The Pacific Ocean thus entered the American imperial imagination in new ways during the postwar period. Familiar commercial imperatives set by the American whale fishery in the antebellum era joined newer more muscular visions of territorial control often wrapped in the untapped and largely unknown environmental potential of places like Alaska.[7]

American naval officers articulated a similar vision of American empire on the Pacific that, in the ideal, emphasized control of the ocean in real and imagined terms. If they agreed on little else, postwar officers such as Captain Robert W. Shufeldt and the war hero Admiral David Dixon Porter were certain that America's future lay in the Pacific. According to naval historian Kenneth Hagan, officers of this generation generally believed that the United States "was inevitably destined to be a great trading nation." Porter, whose father, David, had aggressively shown the American flag in the Pacific during the War of 1812 and who himself had become the postwar navy's most influential officer, believed that "the nations of the earth are looking for the shortest possible route to and from China. The nation that can retain possession of the Eastern trade will be the richest on earth." Shufeldt, meanwhile, declared his belief in 1870 that the Pacific was "the ocean bride of America. It is here . . . upon this sea . . . that the East and West will join hands and the great circle of civilization will be complete. The Pacific Ocean," he concluded, "is and must be essentially American."

But what did that mean exactly? For Shufeldt, Porter, and their allies in Congress, such visions were not just commercial. Rather, they were rooted in a sense that the ocean itself could somehow be controlled and claimed. This control was in part territorial, but it was also rhetorical, imaginative, scientific, and cartographic. In their mind, the Pacific would become a distinctly American place. Shufeldt was not the only one to employ such metaphors. It was a "national duty," the navy's chief hydrographer urged in 1869, to explore, survey, and chart places such as Korea and China as "pioneers on this virgin coast." From Pacific brides to virgin coasts, the ocean seemed to these men ripe for American claims. By rendering the marine environment in gendered terms, such pronouncements also underscored a more masculine, muscular role for the United States and its navy.[8]

Statements such as these also privileged scientific study and cartographic representation as potent tools for extending American control over oceanic environments. "Long stretches of coast must be surveyed," declared Secretary of the Navy George Robeson in 1870, "ports of resort and harbors of refuge on the mainland and in mid ocean must be sounded," and "points of difficulty and of danger tested and marked out." To these hydrographic pursuits, the secretary linked loftier imperial goals. "At vast distances, with thousands of miles between," he continued, "the flag of the republic must be displayed wherever barbarism is ignorant or cupidity unmindful of our rights and power." Robeson thus fused the flag of the Republic to the navy's hydrographic program. Control of natural environments and the Pacific Ocean, in particular, became the work of naval scientists, one arm of American expansion over the sea during the postwar period.[9]

The structure of hydrographic science in the navy changed in response to these postwar realities. In 1866, the navy created the Hydrographic Office, now separate from the observatory, which both reflected the increasing specialization of science as well as the growing complexity and centralization of hydrographic work, which the navy defined as exclusive of astronomy. During this era, the Hydrographic Office served primarily to direct surveys, to supervise chart production, and to disseminate charts to the naval, merchant, and whaling fleets. Since the founding of the Depot of Charts and Instruments in 1830, the navy procured most of its charts through private firms and from the British Admiralty, whose global presence in all oceans marked it as the preeminent sea power of the day. The Royal Navy remained the main hydrographic source for the Americans. By 1885, the Hydrographic Office engraved some 350 charts of its own. It procured and printed approximately three thousand more from the British. Still, in 1879, the navy purchased the engraved plates of E. and G. W. Blunt, an important private supplier of charts to the American maritime community, and secured the copyright to Nathaniel Bowditch's *New American Practical Navigator*. In 1883, the Hydrographic Office once again began to republish Maury's *Wind and Current Charts*, which it had suspended when Maury resigned his officer's commission to join the Confederate Navy in 1861. A year later, it established branch hydrographic offices in American port cities, both on the coast and on inland rivers and lakes. These developments both reflected changes in the postwar maritime world toward inland and coastal maritime commerce and also continued the close relationship between naval science and American commerce. While no cartographic rival to Britannia in terms of sheer quantity of cartographic production, the work of American hydrographers nevertheless was directly tied to American expansion as it grew tentatively over the ocean. It could remain in debt to the Royal Navy for many of its charts and yet still execute important surveys and construct charts of its own that bolstered burgeoning American claims.[10]

For the navy, the postwar era was not one of grand exploring expeditions or new, revolutionary cartographies as the antebellum era had been, but rather of widespread and sustained hydrographic surveying in support of the gradually changing American interests on the ocean described above. No vessel better epitomized the navy's fraught designs for the postwar Pacific than the side-wheel gunboat USS *Saginaw*, which provides a useful lens to understand the routine activities of hydrographic work in the Pacific after the Civil War. In many ways, *Saginaw* represented the nation's designs on the Pacific. Measuring 155 feet in length, drawing only four and half feet of water, and displacing 453 tons, *Saginaw* had been commissioned at the Mare Island Navy Yard in 1860, the first American warship built on the West Coast. In its construction, historian Hans Konrad Van Tilburg shows, Californians invested their youthful Pacific-oriented identity, claiming to eastern critics that they had fashioned a ship out of West Coast materials and resources that were the equal of more established eastern shipyards. The Navy Department had designed *Saginaw* particularly for "service in the China Seas." Like most American warships of that era, it would be a sailing vessel with auxiliary steam engines. Armed originally with one pivot gun forward and two, twenty-four-pound broadside cannons, the gunboat was by no means fashioned to strike terror in enemy fleets. *Saginaw* was neither Nelson's *Victory* nor Fitzroy's *Beagle*; it was built neither to flex muscle nor to accommodate science. It was well-adapted, however, to look after American interests in the Pacific along uncharted coasts with unknown and sometimes dangerous currents and other hazards to navigation. In many ways, the little vessel, considered a third-rate steamer on the navy's rolls, embodied America's incipient aspirations on that ocean.[11]

When *Saginaw* arrived to chart Alaskan waters and protect American interests in its new territory, its commanding officer found a fearsome marine environment in need of the order that charts could provide and that befit ocean waters in transition toward a territory of America's Pacific empire. In April 1868, *Saginaw* departed San Francisco bound for the Alaska Territory to explore and survey and to support the army, whose base at Sitka on the territory's southeast panhandle represented the only political authority of a nation now struggling to govern even its own Southern states. Sitka sat amid the Alexander Archipelago, forming a chain of islands, sounds, and circuitous channels, almost completely uncharted, which today constitute the Alaska Marine Highway. "The line of channel through this myriad of islands is rarely wider than the Hudson," wrote *Saginaw*'s commanding officer, a native New Yorker. "Through these fearful gorges, less than a mile wide, the tide runs with a terrible rapidity, to which our current of Hell Gate, near New York, is a 'mere circumstance.'" *Saginaw* would be the first American warship to navigate and chart these waters. The commodore of the Pacific Squadron ordered *Saginaw* to make "explorations and surveys," and to determine "the most suitable harbors and anchorages on the coast and in the adjacent islands." These might easily have been Secretary of the Navy Paulding's orders to Wilkes nearly a half century before.[12]

Naming, again, proved a powerful way to appropriate new outposts of this ocean empire. As Van Tilburg argues, "the newness of the territory offered . . . opportunities to name or rename natural features in Alaska, to impose a new and familiar template on the wilderness." In early May, for example, *Saginaw* spent two days surveying a harbor known previously as Kake Bay, named after the local Tlingit people, who, like the Fijians of the South Pacific, were known to be the most fearsome in the area. They "paint their faces with red and black paint, giving them a fiendish expression," the captain's clerk, Peveril Meigs, observed in his diary. Kake Bay became Saginaw Bay in the Americans' survey and subsequent charts. After *Saginaw*'s commanding officer, Lieutenant Commander John Mitchell, was mysteriously murdered during a brief return to San Francisco, the ship's new captain, Commander John W. Meade, promptly placed Mitchell Bay on the new chart of Alaskan waters. The finished product, H.O. 225, was "a chart of rough and crude appearance," remarked an Alaskan geographer in 1902, but one "which has been very useful."[13]

*Saginaw*'s men were captivated by Alaskan nature, which seemed to dwarf the little vessel, a defining characteristic of the environment and the resources the Americans had been charged to survey. The hazardous channels formed "one mass of foam," as Meigs put it, and he referred to a "beastly reef," which nearly proved *Saginaw*'s destruction in January 1869. Commander Meade observed great depths of water—"two hundred fathoms of line find no bottom"—and channels that "bewilder the navigator" in an "endless labyrinth." The waters teemed with salmon, and, at least according to Meade, the Russians had not completely extinguished the sea otter whose pelts had been central to Russia's Alaskan empire. *Saginaw* spent considerable time identifying veins of coal useful to bituminous-burning steamers ranging far from the navy's closest coal depot. Alaska seemed to hold the prospect of both danger and bountiful resources.[14]

Meade himself had thought Alaska "very much over-rated," when he took command of *Saginaw* that January, but it appears that by the end of the cruise, he experienced a change of heart. In two articles on that "little known" territory published by *Appleton's Journal* in 1871, Meade wrote of "fish" in "inexhaustible quantity," copper, iron, and silver. Gold, he added, "has been discovered on the Stachine [*sic*] River." He thought that these resources would "soon lead to a considerable development" and "the emigration of our people" to be "permanently watched over by a military and naval force." As yet, the American presence on land and offshore was small, and Meade half-joked about "the progress of American civilization and republican institutions" there. The Alaskan wilderness was a borderland in transition. Meigs deemed it "a dense, damp Russian American forest." In making sense of the Alaskan environment in these ways, he perhaps rendered it in the same transitional and transitory nature that marked the transfer of power from the Russians to the Americans. Years later, in 1923, when Rear Admiral Seaton Schroeder looked back at his time as a midshipman aboard *Saginaw*, he could see the culmination of a half century that had begun with the navy's work in these waters. Alaska's scenery, he remarked, "should make a pleasant theme for the summer people who have taken to visiting . . . in increasing numbers during recent years." Like Fiji and Hawaii, tourism and recreation would ultimately follow American commercial and naval power in the Pacific.[15]

In 1869, however, Alaska remained a dangerous, largely unknown place, a reality borne out when Kake Indians killed two American traders for whom the *Saginaw* now sought retribution. The so-called Kake War continued the tradition of American gunboat diplomacy that the navy had established in the Pacific beginning in 1813 when David Porter's *Essex* arrived at the Marquesas Islands and immediately interceded in a conflict between warring indigenous groups there. In some ways, the Kake War also had parallels in Wilkes's burning of Malolo in 1840. The bodies of the two traders had been found on the beach at what came to be called Murder Cove. A few sinews of flesh remained on their bones. The Americans held the Kake responsible. It seemed to fit their fearsome reputation as scalpers and beheaders living in tense coexistence with white traders and the new imperial power. If the Kakes did not eat people as the Fijians apparently had, Schroeder still referred to them as "these savages." They had killed the two men, ostensibly in retaliation for the death of one of their own at the hands of the American garrison at Kiska on New Year's Day. On February 14, *Saginaw* entered its namesake bay and shelled two abandoned Kake settlements—"spreading the splinters in every direction," Meigs noted. The next day it leveled a third village. "There was but little trouble after that," Schroeder remembered. Such "prompt retribution" prevented the Kake from again "committing excesses." Schroeder's assessment echoed Wilkes's at Malolo. The purpose was much the same. Once again, cartographic transformations such as the one that had replaced the toponym Kake Bay—signifying, as it did, the presence of indigenous meaning—with Saginaw Bay and thus the new American presence, closely preceded acts of violence that consummated the transformation in fire and destruction. Perhaps Rear Admiral Thomas T. Craven, commodore of the Pacific Squadron, recalled his own unhappy days as a young officer in the Ex. Ex. under Wilkes's heavy hand when he received the report of *Saginaw*'s actions from Meade. Then again, perhaps he did not think twice about a policy of retribution against the indigenous peoples of the Pacific that had a deep history in the American navy. Once again, a voyage of exploration and survey had turned to naval power to protect American interests.[16]

In February 1870, *Saginaw* sailed to a different corner of Seward's empire to survey and clear a coral bar blocking the approaches to the inner lagoon at Midway Atoll. Once called Brooks Island for the American shipmaster who first claimed it under the Guano Act of 1856, the atoll became Midway when Captain William Reynolds arrived aboard the steamer *Lackawanna* in 1867. Reynolds, the Ex. Ex. veteran, raised the American flag, and put his surveying experience to work once again to chart the islands and surrounding reefs. Midway had no guano, but it did have a sheltered lagoon and, as its new name attested, a central location for steamers of the Pacific Mail Steamship Company operating under a subsidy from the U.S.

government to run mail and passenger service between California and China along Maury's Great Circle Route. Reynolds, sensing Midway's potential importance, named the inner lagoon Welles Harbor after Secretary of the Navy Gideon Welles and Seward Roads outside its mouth in honor of the secretary of state whose vision had done so much to extend American interests into the North Pacific. Reynolds, who had once been so taken by Cape Horn's Pacific vistas, felt passionately about Midway, too. He was perhaps overly optimistic when he wrote to Welles that Midway possessed "a perfectly secure harbor. The bar at the entrance . . . might be deepened at a very small expense, and a port vastly superior to Honolulu be thus opened to mariners, where a depot might be established for the supply of provisions, water, and fuel."[17] While it is hard to imagine Midway being in any way superior to Honolulu—the latter, Reynolds had helped survey while with the Ex. Ex. in 1841—such rhetoric outlining the atoll's commercial potential piqued interest in the Navy Department.

*Saginaw*'s work at Midway reflected new relationships with the sea, mediated by science, but also by a faith in the destructive (or perhaps constructive?) power of technology to materially change the marine environment. When Reynolds and his fellow surveyors charted the Fijis in 1840, they sought channels through reefs, places in which nature had provided security from its own dangers. Now the Americans intended, in part, to remake the environment where possible. Their instrument would be a new rock-drilling machine, invented and patented by George W. Townsend, the engineer contracted to clear Midway's channel. Originally designed for improvements to the sheltered waters of Boston Harbor, Townsend's machine sat on four legs fully submerged beneath the surface with a central arm to plunge canisters of explosive charges into the coral heads. A diver supervised the work on the bar aided by a boat and a skiff on the surface to raise and remove the large severed heads of coral. *Saginaw* arrived at Midway in March 1870 with two hundred kegs of blasting powder, the drill, and the deep sea diving rig soon after. Townsend and his men represented the technological progress of the age. They were "Yankee experts in submarine diving and blasting," according to one sympathetic writer. "We now have an opportunity to see a coral reef attacked," the *Hawaiian Gazette* anticipated with a barely concealed hint of militaristic conquest over nature.[18]

Where once the reef symbolized the navigator's worst fears, during the postwar era it seemed altogether more nuisance than existential threat. Reefs had long been implacable hazards, but now new technologies held the possibility of obliterating the reef itself. Captain Charles Wolcott Brooks, who had been a great promotor of Midway since he first laid claim to it in 1859, remembered his "prospecting cruise" in an 1870 promotional article. "The ocean in this vicinity, and the chain of islands to be visited, had been very imperfectly explored, and for more than half a century its dangers had proved a great bug-bear to many whalers, and, to some, a final resting-place." To Brooks, these islands and atolls promised "unlimited opportunity for enterprise, requiring but capital and skill, well directed, to develop a future our anticipations can scarcely over-estimate. America is not wonting in either of these," he concluded, "she now has this ocean to herself." Perhaps he was getting ahead of himself, but such rhetoric reflected similar beliefs in the navy and in Congress that the Pacific was becoming ever-more an American place. It remained simply to identify the value of various islands and coasts on the chart, Brooks believed, and then to alter them through enterprise and technical ingenuity to fit American needs.[19] Altering the material environment to suit these interests thus became one more aspect of a postwar ocean in transition, bent to the will of American politicians, naval officers, surveyors, and cartographers. To blast away at coral heads while, outside the lagoon, the ocean rolled on in all its vastness, dynamism, and natural power was to begin to chip away at the notion that the ocean continued to seem boundless and wild in the imagination.

Yet for all the Americans' faith in human enterprise and technical ingenuity—for all their visions of an ocean on which the United States might enact its postwar ambitions—the natural world flouted pretensions. Just as Wilkes's men found their trigonometric survey daily undermined by the dynamism and power of the sea, so too did the Americans at Midway three decades later. "The weather has . . . been exceedingly unfavorable to the kind of work we have been engaged in," wrote *Saginaw*'s new commanding officer Lieutenant Commander Montgomery Sicard. He reported to the commodore of the Pacific Squadron of strong gales from the north and east, and alternately, "a long and heavy ground swell" from the west, which "broke across the whole bar, and also across Seward's Roads, in the most furious manner." The "send of the sea," as Sicard put it, made work difficult, and often treacherous, for the civilian diver who sometimes found his own "umbilical cord" and lifeline to the surface snagged around coral heads. The coral itself was hard to blast away. Rather than shattering into large pieces as the Americans hoped and expected, Sicard noted that it was "so soft in places that it will work in the hands like putty." The reef's consistency blunted the charged canister's explosive effect. With currents rendering the coral debris too difficult to bring to the surface, Sicard now hoped that the same currents would carry away the debris by themselves. Here too his hopes were dashed. The marine environment refused to work in the Americans' favor.[20]

Money soon ran out. The spendthrift Reconstruction Congress had originally appropriated $50,000 for the project in 1869. In seven months of work, Saginaw's men and its engineers had succeeded in blasting a channel across the bar measuring fifteen yards across and four hundred in length, not nearly wide enough for the Pacific Mail steamers to navigate. Midway's placid lagoon beckoned, but coral quite literally barred the way. Sicard estimated that an additional fifteen months and a $214,000 appropriation would be required to complete the job. This far exceeded initial estimates. "More difficulty has been experienced and greater obstacles encountered than were anticipated," Secretary Robeson admitted. The Saginaw departed Midway in October 1870. Its mission had largely been a failure. "It really seems a questionable undertaking to attempt the formation of an anchorage here for the large steamers of the Pacific Mail Company," Saginaw's pay inspector George H. Read decided. In an era when new technologies and new forms of destructive power promised to transform not only the ways Americans imagined the sea but also to change it in material ways, the marine environment undermined their efforts.[21]

Indeed, the environment itself perhaps represented the most formidable challenge to American intentions in the Pacific during the last quarter of the nineteenth century. It not only thwarted Saginaw's voyage, it ended it. When Saginaw had departed San Francisco bound for Midway the previous February, Read remarked, "Old Neptune gives us a boisterous welcome to his dominions," reminding the crew that they were "once more his subjects." Here, Read sensed the agency of nature, that he and the crew, remained, to a degree, at the mercy of the natural world. Now, as Saginaw left Midway, one of his shipmates confessed a strange feeling of superstition regarding the coming voyage. "He felt greatly depressed without being able to define any cause for it and that he could not rid himself of the impression that some misfortune was impending. He expressed a firm belief that we should meet with some disaster on our voyage," he remarked. Perhaps Read embellished the moment for dramatic effect when he published his narrative of the voyage some forty years later, but it may also suggest the pervasiveness of the maritime folklore that continued to structure these mariners' worldview even as steam engines and drilling machines changed the nature of ocean voyages, the relationship of mariners to the sea, and the physical nature of the marine environment itself.[22]

On the steamer's return to Honolulu, Sicard intended to visit Ocean Island in Kure Atoll, fifty nautical miles west of Midway, to be sure that the island had been placed correctly on the navy's charts and to rescue any shipwrecked sailors marooned there in this out-of-the-way place—the most remote of all Pacific atolls. On the moonless night of October 29, *Saginaw*, borne westward by an unknown current with more speed than Sicard knew, struck the reef at Kure Atoll, broke apart, and sank. The reef, which *Saginaw* hoped to destroy at Midway, ultimately claimed the ship itself as it went to pieces "in the grip of the breakers," as Read put it. He interpreted the wreck as a battle of man and machine against nature as the ship entered "upon the death-struggle with the rocks. There is no wonder that brave men—men having withstood the shock of battle and endured the hardships of the fiercest storms—should feel their nerves shaken," he observed. The crew remained marooned on the island for more than two months until a boat sent by Sicard made a perilous twelve-hundred-mile open-ocean voyage to seek help from Hawaii. Sicard and Seaman William Halford, the only survivor of the five-man boat crew, were celebrated for feats of endurance in this remote, inhospitable watery world. But *Saginaw* was gone, claimed by the same reefs it had initially been ordered to chart and destroy to make the ocean safe for maritime commerce. Places like Midway and Ocean Island would remain, for now, inconspicuous outposts of empire more treacherous to navigation than secure harbors of refuge for American steamers.[23]

All across Seward's empire, naval science and exploration ventured into extreme environments from remote Ocean Island to the jungles of Central America, the icy Arctic, and the deep sea, confident that natural obstacles could be overcome by human endurance if not by mechanical power. Again and again, the Americans were humbled. Naval and military historians have generally perceived this era of American naval history as a time of professional stagnation, institutional constriction, and technological retrogression. Yet, the 1870s was a decade of extraordinary, diverse, and significant activity for the navy and for naval science specifically. When *Saginaw* wrecked in the North Pacific, the Navy Department had intended, next, to send the vessel to the coast of Panama where an American expedition was exploring routes for a proposed canal. This was part of a larger effort that saw various naval surveyors investigating passages from Panama to Nicaragua and Mexico.

The proposed isthmian canal mirrored similar efforts at Suez in 1869 and emerged from antebellum transits of Forty-Niners by mule or rail across the isthmus seeking the shortest possible journey to California. Replacing these cumbersome modes of transportation with an all-water route would more closely bring together the American coasts and open new highways of maritime commerce between oceans. From the beginning, such efforts had been primarily commercial in nature, though the strategic value of any canal to the United States and its navy was also apparent. Nevertheless, the isthmian canal would not take on its more explicit strategic importance to the navy until the end of the nineteenth century. Still, such a feat amid the malaria and yellow fever–infested jungles would serve a greater purpose. As Secretary Robeson put it in 1869, the "value of such a work" was "in its effect upon commerce, and, through commerce, upon civilization throughout the world. The time has come for action in the field," he declared. Commander Thomas O. Selfridge Jr., who led a number of these surveys in the early 1870s, regarded the canal as "the grandest triumph of this or any other age." Of course, it had not yet been built. Yet such aspirations joined a growing chorus of voices who increasingly saw national destiny in the Pacific in commercial, if not yet strategic, terms. The navy and its surveyors and cartographers went to work, building on the commercial empire ushered by the Ex. Ex., by Maury's *Wind and Current Charts*, and by the American maritime world more generally. To be sure, the American merchant marine was a shell of its antebellum self. The navy too had been much decreased from its Civil War numbers. Its technological advantages in steam power, ironclad construction, and revolving turrets went by the wayside, but if we look at the many unheralded ventures of naval surveyors and cartographers in the era between the Civil War and the War with Spain in 1898, a slow, but discernible evolution in the changing nature of American ocean empire emerges.[24]

The navy's postwar canal surveys were not primarily hydrographic; rather, they involved overland exploration and mapping, but even as naval surveyors and engineers proposed a complex system of locks and tunnels on land, they surveyed the Central American coast for harbors to protect and facilitate this interocean highway. Like Midway, various surveyors proposed changes to the isthmian littoral commensurate with the engineering projects proposed inland. In 1873, Selfridge reported to Secretary Robeson on the mouth of the Atrato River on the Isthmus of Darien, Province of Panama. He noted the "magnificent harbor at the mouth of the Atrato, named by the expedition Columbia Harbor." He went on to tell the secretary that the body of water was "ten miles deep by five wide, with a uniform depth of ten fathoms, completely land-locked and easy of access, it has no superior," Selfridge concluded.[25]

Farther north, at Greytown in Nicaragua, Commander Edward P. Lull reported to the secretary on the natural advantages and challenges of that route. Silting at the harbor's mouth was particularly problematic. "There are islands where twenty years ago there was water enough to float a frigate," he observed. For Lull, the process of environmental change along the Nicaraguan coast would be a decisive factor in the construction of any future canal there. American ingenuity would again need to be brought to bear to overcome such environmental obstacles. Greytown's harbor was "commodious," but "the question is, can the harbor be restored?" Lull wondered. He immediately sent "a quantity of sand" to the Smithsonian, where Superintendent Joseph Henry tested it and reported to Lull that it had not originated from the sea; it had been deposited by the San Juan River. Turning to the engineering and technological potential that suffused the entire canal project, Lull proposed to divert the San Juan River through a series of canals, dredge the harbor, and construct a breakwater. "After which there will be nothing to again destroy it," he concluded. In 1873, the Panama Canal remained nearly forty years in the future. The delay, of course, had much to do with the environment. From the Cordillera Mountain Range inland, to the malarial coasts, the hydrographic challenges offshore, and, most infamously, an erupting volcano along the Nicaragua route, naval surveyors weighed natural advantages and disadvantages within the broader political, diplomatic, and technical contexts of building an isthmian canal. Once again, technology and engineering could not yet overcome these natural obstacles.[26]

Even as Selfridge and Lull sweated in the tropics, the navy ventured into Baffin Bay, the Bering Sea, and the Arctic Circle, hoping to discover an open polar ocean rumored in Maury's day and rooted in the same polar milieu as John Cleves Symmes and Jeremiah N. Reynolds, whose theories had drawn the Ex. Ex. toward a different pole in the 1830s. The discovery of this polar sea might lead to leaps in hydrographic understanding, scientific laurels, and commercial promise. But, as historians Michael Robinson and Aaron Sachs have argued, Arctic exploration in this era was also about other, more intangible, but no less powerful ideals such as bravery, rugged masculinity, and an American cultural claim to these icy regions. As Robinson argues, "stories, more than specimens or scientific observations, constituted the real currency of Arctic exploration." The Arctic, he continues, could be "explored without being administered, a place to flex imperial muscle without having to do the heavy lifting required by a colonial empire." Robinson refers to this sort of expansion as America's "imperial

HOW TO DISPOSE OF THE REMAINS OF OUR NAVY.

*How to Dispose of the Remains of Our Navy*, a lithograph by Mayer, Merkel, and Ottmann, appearing in *Puck*, 1882. It shows the ill-fated *Jeanette* Expedition and the numerous vessels the navy sent, each in search of its predecessor, amid a whimsical, half-comical, half-terrifying Arctic environment. Courtesy of Boston Public Library.

pubescence." Indeed, we might see such awkward sometimes flagrant, sometimes stifled—displays of American power throughout the Pacific and in the rhetoric of naval officers and politicians alike during this period.[27]

The navy's encounter with the Arctic was generally disastrous, consisting of public–private ventures, sometimes poorly led, all confronting icy seas that both spurred the public spectacle of Arctic voyaging and, sometimes, claimed the voyagers' lives in the process. From the Grinnell expeditions of the 1850s to the *Polaris* and *Jeannette* expeditions of the 1870s, many ended in failure if not the deaths of its members. In 1882, a cartoon on the back cover of the satirical magazine *Puck* summed up more than a decade of naval exploration to the pole by showing a retinue of shattered vessels precariously wrecked among icebergs and polar bears, each ship sent out to rescue its foundering predecessor. The image is underscored by the title, "How to Dispose of the Remains of our Navy," poking fun at the state of the postwar fleet and the follies of Arctic exploration. In its title and iconography, the image recalls the British painter Edward Landseer's fanciful *Man Proposes, God Disposes* in which two voracious polar bears make a last meal out of the remains of Sir John Franklin's ill-fated expedition to the Arctic in 1845. As Robinson contends, these voyages were as much if not more about the extreme environment—about human perseverance in the face of an inhospitable nature—as they were about science. In their death throes, they also produced compelling narratives and newspaper headlines with "sensational and alarming reports" of survival, infighting, alcoholism, and even poisoning. Ironically enough, cannibalism was once again often rumored and sometimes did take place, not among the Eskimo and Inuit peoples of the Arctic who did so much to aid these Americans, but among the desperate, half-starved white men themselves. These efforts were also tense partnerships among private investors, merchant or whaling masters, civilian scientists, and the navy—all with a stake in the journey, but with little else in common. No wonder such undertakings were exceedingly difficult to execute successfully.[28]

Charting and studying this environment became secondary to simply surviving it and living to tell the tale in widely read narratives. Charles Francis Hall, an American whaling captain and amateur scientist, led the *Polaris* Expedition in 1871, which the navy fitted out and supervised, but which was otherwise indicative of this strange cast of Arctic American characters. "Polaris," which was the name of Hall's ship, "and Newman's Bays were surveyed, and the coast-line to the southward of Polaris Bay was examined for over seventy miles." Robeson must have been tickled when his name appeared on the chart gracing a newly discovered strait. He could add himself to the long list of navy secretaries whose surnames dotted the world's oceans and the Americans' charts. Hall shrewdly named a never-before-seen form of bottom-dwelling organism *Protobathybins robesonii*, a classification that, for a number of possible reasons, has not stuck. Indeed, the navy secretary was at first heartened at the "expectation and . . . hope of large and valuable additions to the domain of human knowledge." Spencer F. Baird, Joseph Henry's colleague at the Smithsonian and "the distinguished scientist," as Robeson called him, had apparently "taken a great interest in the expedition," making recommendations about the scientific questions to be studied and about the instruments with which to answer them.[29]

Yet the Arctic environment soon overcame the Polaris Expedition. Morale devolved. Hall died under suspicious circumstances. *Polaris* was crushed by ice and sank. The remaining men barely returned to the United States alive. Their chart of the Arctic, replete with names such as Repulse Harbor, Thank God Harbor, and Providence Berg, were enough to tell the harrowing story. Maury Sea seemed to hold the potential of the elusive polar ocean, which had impelled so many expeditions to the Arctic in the first place, but the Americans had been turned back before further investigation despite their intention to "honor our dear flag, and to hoist her on the most northern part of the earth." The *Polaris* Expedition had largely been a failure. "Unfortunately there was not much opportunity for taking soundings," one of the survivors testified in a naval court of inquiry, which consisted, among other members, of Robeson, William Reynolds, now a rear admiral, and the scientist Baird. Baird's friend and Maury's old nemesis, National Academy of Science president Joseph Henry concluded, "great difficulty was met with in obtaining men of the proper scientific acquirements to embark in an enterprise which must necessarily be attended with much privation, and in which, in a measure, science must be subordinate." The Arctic environment claimed lives, ships, and with them—often—the scientific agendas of these expeditions. As Henry noted, science was subordinated to survival in extreme environments where conquest over nature or near-death survival narratives tested the nation's physical endurance writ large and satiated a public appetite for harrowing stories of exploration and intrigue in remote, far-off seas.[30]

For the navy, Arctic exploration was a way to stake an American claim to high latitudes as the Ex. Ex. had done to Antarctica and as a dwindling commercial whaling fleet worked its way farther into the frigid Kamchatka and Bering seas. The most infamous effort was the *Jeannette* Expedition, 1879 to 1882, which ended in the deaths of twelve men, including the expedition's leader, Lieutenant Commander George W. De Long, when his ship *Jeannette* became hopelessly frozen in ice near Russian Siberia. In 1882, the navy's hydrographer, John C. P. de Krafft, offered a fitting epitaph to the expedition's significance. "Their hazardous mission was pursued fearlessly, and with great zeal and energy. Although the results achieved are not commensurate with the hardships and loss of life endured by these bold workers in the cause of scientific discovery, yet they are nevertheless very important." He then went on to list the ill-fated expedition's hydrographic contributions. Besides confirming theories of Arctic ice circulation derived, ironically, from the drift of *Jeanette*'s wreckage around the pole, the expedition had discovered Jeannette, Henrietta, and Bennett Islands in what came to be known as the De Long Group. This, de Krafft said with a melancholy flourish, was "a lasting testimonial in the regions of eternal ice to the intrepidity of the commander . . . who with indomitable energy reached a higher latitude than any heretofore attained in the Siberian Arctic, and gave his life to the cause of scientific research, while inscribing his name for ages where few may hope to follow." From Midway to the isthmus and the Arctic, nature continued to flout scientific pursuits and the navy's broader efforts to conquer marine environments through science and feats of physical endurance. As environmental historian Aaron Sachs contends, referring to De Long and to *Jeanette*'s chief engineer, George W. Melville, these men seemed to set out "with dreams of conquering nature." De Long never returned. Melville, who survived to come home a chastened hero, was "not the conqueror of new lands." His was "the heroism of endurance and adaptation," Sachs writes, "of respect for the forces of nature, of acceptance."[31]

The navy achieved more success in deep sea exploration, breaching an environment even more inaccessible and inhospitable to study than the Arctic, but one in which technology and ingenuity promised new discoveries for the United States. When USS *Tuscarora* departed San Francisco in 1873 and set off across the Pacific to chart a course for a trans-Pacific submarine telegraph cable, it was bound not for Arctic poles, but for the deepest recesses of the ocean floor. Humans could not go there, of course, but machines could, and they might bring up a sample of the bottom of interest to scientists as well. As ship's writer Henry Cummings made clear at the beginning of this chapter, *Tuscarora*'s Thomson Deep Sea Sounding Machine was a technological spectacle, a machine that the Americans would come to regard with something close to awe, imbued with a particular power to transform understandings of the ocean. This was the era of the Second Industrial Revolution. The Philadelphia Centennial Exposition with its massive Corliss steam engine was less than three years distant. Technology seemed to represent all the progress of the age and of America's growing ambitions on the world stage. Likewise, as steam power eclipsed canvas sails and as steel vessels gradually replaced wooden and iron-hulled ships in the 1880s and 1890s, the navy's relationship with the marine environment fundamentally changed, too. As one naval officer remembered it, "From the contests of old days with wind and weather, when the battle between elements and man was always uncertain, [the author] has lived to see the final triumph of man in these high-powered vessels of today, which defy

winds and weather and run almost as railway trains on schedule time."
Reefs and ice still presented hazards. Ships remained, to a degree, at the
mercy of currents and storms, but the dangers of the natural world—perhaps
contrary to the experience of Sicard at Midway, and Hall and De Long in
the Arctic—generally seemed to be mitigated by modern technology.[32]

Naval surveyors and hydrographers saw in machines the possibility to
transform the study of the ocean and to reach dimensions of the marine environment that were previously inaccessible. While technology could still
be humbled before nature's power, the 1870s were a transitional moment
for naval scientists' conception of the ocean they sought to study. Ideas of
ocean wilderness that had served nineteenth-century hydrographers as a
foil to their own work did not hold as much significance in this era of technological change that culminated in the new steam and steel navy of the
turn-of-the-century. The mysteries with which Herman Melville infused his
ship-destroying white whale were fading. In the futuristic writing of Jules
Verne, who published *Twenty Thousand Leagues under the Sea* in 1870, the
ship itself—Nemo's *Nautilus*—was the very sea monster that Moby Dick had
once been. Now it was mechanical and obeyed human control. Not coincidentally, Verne kept a copy of Maury's *The Physical Geography of the Sea*
next to him as he wrote *Twenty Thousand Leagues*, trying to imagine and to
navigate a submarine world that remained fanciful, but increasingly within
the realm of scientific exploration, discovery, and study.

Currents of change raced through the maritime world as far off lands appeared to come under the perceived civilizing influence of Euro-American
empire. By the early twentieth century, one writer could muse that in the
Fijis, which Britain had colonized in 1874, "a white man or woman is safer
with these natives than on the streets of New York or Chicago." Traveling in
those same islands in 1895, Mark Twain quipped, "Sixty years ago they were
sunk in darkness . . . now they have the bicycle." When the old Mississippi
River pilot heard of the death of Charles Wilkes's wife, he mused, "[Wilkes]
had gone wandering about the globe in his ships and had looked with his
own eyes upon its furthest corners, its dreamlands—names and places which
existed rather as shadows and rumors than as realities." Here, Twain seemed
to refer to a bygone wilderness, fading before the progressive advance of
humans and their technology. "But everybody visits those places now in out-
ings and summer excursions," he concluded, "and no fame is to be gotten out
of it." By the 1870s, the sea and the study of the marine environment were
changing, or at least hydrographers and the American public perceived that
some fundamental shift had occurred. The deep open sea had become a

destination, itself a site for scientific study. As historian of science Helen Rozwadowski has written, "between 1840 and 1880, the ocean ceased being a wasteland and highway and was transformed into a destination, a frontier, an uncivilized place ripe for conquest and exploitation."[33]

For the navy, the deep sea held the promise not only of new scientific discoveries but also of commercial and imperial value as submarine telegraph cable networks began to knit together continents in nearly instantaneous communication across vast expanses of water. It was increasingly this latter objective more than questions of pure science that naval hydrographers pursued in the postwar era. Indeed, as early as the antebellum era, naval officers like Wilkes looked askance at the scientific work of his corps of naturalists, preferring instead to focus on questions of navigation or hydrography's practical ends. As one newspaper put it, *Tuscarora's* work was "most practically useful. It is no mere idle curiosity that has to be satisfied," a thinly veiled critique of theoretical science. A proposed cable connecting Japan to California and points in between was the impetus for *Tuscarora's* voyage. It required the kind of deep sea sounding that Maury had directed in the laying of the Atlantic cable. The Pacific, however, presented a different scale altogether. After calling at Panama to aid Commander Selfridge in his survey of the isthmus, *Tuscarora* returned to San Francisco and then set off for Alaska under the command of Commander George E. Belknap in September 1873. Belknap was an odd choice. He had none of Wilkes's experience. He was not a hydrographic inventor like John M. Brooke. He had little expertise in hydrographic science. He wrote that his orders "were unexpected and the duty [was] entirely different from the routine of ordinary cruising. No one on board had had any experience in deep sea work, but had heard much of its difficulties," and, he noted, no one at Mare Island was particularly "anxious to volunteer for such work." The navy remained generally unfriendly to science, but Belknap was a quick study. Not unlike the old wooden steam and sail *Tuscarora* with the strange Thomson sounder perched on its gangway, Belknap melded the study of deep sea sounding, his new mechanical sounder, and the relative virtues of piano sounding wire over hemp rope to his own expertise as a naval officer, so that by the end of his career, he was known as a naval scientist among the ranks of Wilkes, Maury, and Brooke.[34]

Belknap was quite evidently fascinated by the Thomson Deep Sea Sounding Machine, reveling in the mechanism. He corresponded with its inventor, the British hydrographer Sir William Thomson, and with the old naval surveyor and inventor Brooke. He made his own adjustments to the apparatus—replacing hemp rope with steel piano wire, adding a detachable

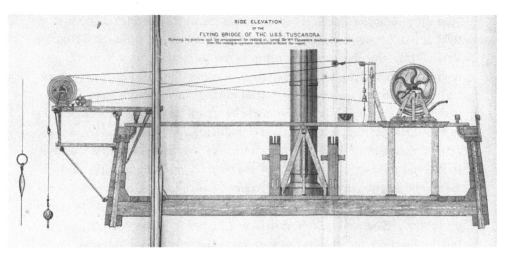

Thomson Deep Sea Sounder, USS *Tuscarora*, 1873–74, showing the configuration of the mechanism atop the flying bridge that Belknap and his men found so marvelous, from *Deep-Sea Soundings in the North Pacific*. Courtesy of Sterling Memorial Library, Yale University.

weight for taking bottom samples much like Brooke's antebellum sounder, and cups for measuring water temperature at various depths. The primary advance of Thomson's design was its dynamometer, which applied tension to the wire so that the surveyor would know exactly when the weight had reached bottom. This had been the vexing question of deep sea sounding from the beginning as the surveyor continued to count fathoms while submarine currents sometimes took the sounder horizontally rather than at a right angle from the sea floor. Against the extreme challenges presented by the deep sea, Belknap wrote, the machine "met every requirement, whether for accuracy of results or celerity of accomplishment." He admired it as a spectacle of this technological age. "One never tired watching the working of the reel," Belknap mused, "so noiseless and perfect in its action. This might be called the very poetry of deep-sea sounding, as, indeed, it was its very perfection." Belknap's remarks are illustrative of what historian of technology John Kasson called "the aesthetics of the machine," which clearly affected Belknap and his men in the beauty of its operation and the wonders of the unseen natural world that it promised to reveal.[35]

Each successful cast seemed a triumph of the sounder as it brought *Tuscarora*'s crew in virtual contact with the little-known, remote world of the deep sea floor. As historians Keith Benson, Helen Rozwadowski, and David Van Keuren have written with a nod to Leo Marx's *The Machine in the Garden*, "the inter-relationship between the machine and Neptune's garden is as tight as that between the railroad and the picturesque yet industrial Hudson Valley." Like the chart's relationship to the terrestrial map, hydrographic technology could not alter the marine environment in the ways technology on land could engineer terrestrial environments into a so-called Second Nature, that is, landscapes remade by technology and human intention to reflect larger social, commercial, political, and cultural aims. At sea, rather, technology, along with the chart itself, became one of the few tools that could mediate hydrographers' encounters with the marine environment, and so they invested it with tremendous imaginative power. "Every cast and answering demand on the apparatus seemed a revelation," Belknap exclaimed. "The lead started down on its mission to the dark abysms below to question and snatch answers from the mysteries hidden since the 'gathering together of the waters.'" *Tuscarora*'s executive officer, Belknap's second-in-command, Lieutenant Commander Theodore F. Jewell, "felt the shock of striking at eighteen hundred fathoms." Navigators had been using sounding line for millennia, feeling their way across the sea floor, but here Jewell was struck by touching bottom through the medium of the sounding apparatus. When he remarked, "the bed of the ocean lies but at our hand," it seems he was speaking both literally and metaphorically. The deep sea sounder thus became an important instrument for the material study of the ocean environment, retrieving samples of the ocean floor to be returned to the United States for scientific study and registering depth soundings for the submarine cable, while also conjuring feelings of exhilaration among these hydrographers, not only in the workings of the mechanism, but also in the previously inaccessible places to which it seemed to convey them imaginatively.[36]

*Tuscarora*'s voyage, like that of *Saginaw* a decade earlier, knit together in voyages of hydrographic surveying a Pacific Ocean increasingly of strategic interest to the United States. From Alaska, it returned to California before setting off in the winter of 1874 across the North Pacific to Hawaii, the Bonin Islands, and Japan. It returned to the United States by way of Kamchatka and the Aleutians, reaching San Francisco once again in September 1874. The Americans had sailed and steamed over 16,600 miles of ocean and made 483 casts, each recorded in minute detail according to its time, location, temperature at various depths, and the material brought up by the sounder from the sea floor variously marked as mud, sand, pebbles, gravel, shells, or ooze. In addition to new discoveries of currents and ocean temperatures, the expedition's chief contribution to knowledge was the discovery of trenches and ridges all over the Pacific Basin. The ocean floor did not simply decline on a plain to its center and rise again near its land peripheries. Belknap reported to the Navy Department that a telegraph cable might successfully be laid on both a northern and a central route, though each had its advantages and disadvantages. North of Japan, the Americans cast the deepest sounding to that time at 4,655 fathoms—more than five miles down—and even deeper than the contemporaneous British-led Challenger Expedition's soundings in the newly discovered Marianas Trench. Like Wilkes's pursuit of Cook's Ne Plus Ultra or Hall's quest for the North Pole, this deep sea work was as much nationalistic as scientific and commercial. Deepest cast, like farthest north or south, became wrapped in scientific rivalry that spoke as much to contests of knowledge as emerging imperial claims in the Pacific. The 1870s were a continuation of an expansive impulse at sea that accorded national glory to the exploration of environments that were ever more remote, inaccessible, and inhospitable to human life. Though the British contested the Americans' claim, as Sir James Ross had of Wilkes's discovery of the Antarctic continent in 1840, Belknap could afford to be conciliatory. "No doubt has been thrown upon the 'Challenger's' admirable work by any American pen the writer has ever heard of," he retorted. Indeed, such deep soundings had become routine for the Americans, who commonly cast in depths over four thousand fathoms. They had become "so expert," Belknap bragged, that anything less than the four thousand fathom threshold seemed "a light matter." Joviality reigned aboard *Tuscarora*. The crew was flush with accomplishment. The officers "would gather in the port gangway with a postprandial cigar, to watch the reel, speculate as to the depth, and possibly indulge in a little quiet betting," Belknap told his readers. In improving on the Briton Thomson's design and casting, the deepest sounding in the history of deep sea hydrography, the Americans had staked a claim not only to the technical mastery of their machine but to the expertise in using it, and, not coincidentally, to the deep sea floor itself by successfully reaching its deepest known recesses.[37]

Like Maury's antebellum work on the North Atlantic telegraph plateau, *Tuscarora's* postwar soundings further revealed the world of the deep ocean, brought it into American consciousness, and suggested that the ocean's greatest depths had been pierced. Rozwadowski argues, "when the ocean was envisioned as a new frontier, routine hydrographic cruises, especially deep-sea sounding cruises, were heralded as national achievements." With headlines such as "Secrets of the Deep" and "Deep Sea Researches Rendered

Easy," the newspapers of the day referred to "the general interest shown toward all that is connected with the series of deep-sea soundings in the Pacific" and work that was "of very great interest to the public." Belknap and his men did much to contribute to these changing popular understandings through published narratives of the voyage. "A retreat of the waters would show a continent buttressed and bastioned like an immense fortress," Belknap wrote, framing the marine environment in military terms that foreshadowed the militarization of the marine environment by naval science at the end of the century, which I examine in chapters 5 and 6. The American press reflected and perpetuated such conceptions and metaphorical devices. One unidentified correspondent, who had attended Belknap's detailed address on the Thomson sounder while *Tuscarora* lay at anchor in Yokohoma, Japan, told readers that naval scientists "are to-day able to give a map of the bottom of the Pacific stretching over thousands of miles. Just think of it," he exclaimed, "a map of the bottom of the ocean! Hills and valleys, river beds and water courses, and even vast mountains, are unearthed, or, more properly, unoceaned by these investigations. The poor mermaid becomes even a greater myth than before, when Commander Belknap's sounding cups disclose the fact that the ocean bed isn't very well adapted to gardening, and even poor old Daddy Neptune slinks away in the dim distance as we reflect upon the surmise that 20 degrees Fahrenheit is not well adapted to his style of dress." Myth and folklore appeared to recede before the advance of naval science. The ocean seemed open to inquiry, less dangerous and not nearly as mysterious. Such sentiments echoed the words of *Tuscarora's* Henry Cummings, whose observations began this chapter. "The scientific world . . . is indebted to us for the light thrown upon a subject hitherto enveloped in mystery and conjecture," Cummings concluded in his narrative of the voyage, which he published almost immediately after the ship returned to San Francisco. "It need be a source of little surprise, if, at a not very distant day, not only the boundaries of continents, countries and states, the location of empires, kingdoms, cities and towns be taught to the rising generation, but submarine geography also be equally familiar to the student and schoolboy."[38] For Cummings, Belknap, and others, the deep sea floor thus entered the American imagination through hydrographic charts just as maps and school geographies framed empires and kingdoms in the imagination of schoolchildren. Rather than the kingdoms and empires of old, however, this new sea floor seemed the purview of Americans by right of discovery, even as the British made their own claims in deep sea sounding.

*Tuscarora's* voyage and, indeed, naval science in general had much to contribute to larger scientific understandings about the marine environment in this era. Even as scientific elites such as Joseph Henry and Spencer Baird perhaps looked askance, if not with disinterest, at American adventurism in the Arctic, they continued to see in the navy and in naval science a collaborative, if awkward, partner in the pursuit of knowledge. One civilian naturalist accompanied *Tuscarora* on its voyage, and the navy sent the deep sea specimens to the Smithsonian to be studied and analyzed. This was no grand, awkward marriage of civilian and military science on the scale of the Ex. Ex., but neither was it a divorce. Belknap himself acknowledged mutual interest, citing the celebrated British naturalist Sir Wyville Thompson when he remarked, "the land of promise for the naturalist was the bottom of the sea." Indeed, Thompson led the scientific corps attached to the Royal Navy's Challenger Expedition from 1872 to 1876, which marked the emergence of the field of oceanography, subsuming hydrography within a much broader inquiry into the marine environment that encompassed physics, chemistry, geology, and biology. If the ocean environment remained a difficult laboratory for urbane, ivory-tower civilian scientists, the navy could help to take them there, or at least return home with specimens of little direct value to the navy but of considerable interest to the scientific world.[39]

Nowhere was the relationship between naval and civilian science more vital and collaborative during this period than in the Coast and Geodetic Survey, where the naval officer Charles D. Sigsbee worked with the eminent naturalist-engineer Alexander Agassiz aboard the steamer *Blake* in the Gulf of Mexico and the Caribbean Sea from 1877 to 1878. Since its inception in 1807, the Coast Survey had employed naval officers to command and man its ships and surveys. In doing so, it provided an outlet for naval officers in time of peace and served as a fertile ground for the growth of scientifically minded officers. Of course, the two institutions competed as much as they cooperated, but for all Maury's disputes with Alexander Dallas Bache, American scientists also held naval scientists such as the astronomers James M. Gillis and Charles H. Davis in high esteem. Like Belknap, Sigsbee was a technologist. He was fascinated with understanding and improving Thomson's deep sea sounder, as he put it, "from a mechanical standpoint." He did so primarily by improving Brooke's detacher, by devising an accumulator—or rubber band—to lessen the strain on the line as the vessel pitched and rolled in rough seas, and by attaching a steam engine to the reel to more easily haul it in. He also invented cylinders that he attached to Agassiz's dredges to capture marine organisms. Sigsbee's adaptations became the standard in deep sea sounding for the next half century. The Navy Department appointed him hydrographer from 1893 to 1897. He then went on to captain the ill-fated *Maine*, a command for which he is better known to history. Nevertheless, his sounder achieved record depths in the Gulf of Mexico in what came to be known as the Sigsbee Deep. More important, for scientists such as Agassiz, the Sigsbee sounder, along with *Blake*'s dredges, brought to the surface new and fascinating forms of marine life and bottom sediment to study. As Sigsbee put it in a letter to Belknap describing his adaptations to the sounder, "the scientific 'fellers' are curious."[40]

If the *Blake*'s cruise evinced cooperation between Sigsbee and Agassiz, it also revealed widening differences between naval and civilian science, which nevertheless coalesced in a common interest in deep sea technology and the results it promised for naval officer and scientist alike. The survey's superintendent, Carlile P. Patterson, reflected the increasing professionalization of the sciences in the late nineteenth century when he observed that "naval officers are professionally neither naturalists nor geologists." Agassiz echoed Patterson, referring to the *Blake*'s "novel work, so foreign to [the naval officer's] usual routine." Sigsbee complained to Belknap, meanwhile, that the Navy Department seemed neither to care nor to publicize the work done by officers assigned to the survey. If naval officers thought much about marine science at all, they continued to be interested primarily in hydrography, which had utilitarian and applied value for their own profession. Scientists in the emerging field of oceanography, meanwhile, quite literally cast their net widely and were interested in theoretical questions as much, if not more than applied science. Yet despite diverging interests, Sigsbee and Agassiz remained connected by mutual interest in the machine, which both registered depths for the chart and brought up samples for study. It was on this common ground, it seemed, that the two men were able to collaborate. Naval officers "were ever ready to promote the special interests during my connection with the vessel," Agassiz observed. Meanwhile, Sigsbee deemed the association "short," but "very pleasant. Professor Agassiz and I were shipmates and messmates . . . during which time we gave ourselves up heartily to the work at hand. To these gentlemen belongs all the credit, pertaining to the province of the naturalist, which was gained on the dredging cruises."[41] The professional working relationship that emerged between Sigsbee and Agassiz during *Blake*'s voyage transcended a widening professional schism between scientifically inclined naval officers and academically trained scientists. By the late nineteenth century, their cooperative efforts were strained. Each looked at the other as an outsider, whose differences could yet be resolved in the potential of the deep sea sounder.

Yet even as naval and civilian science found common ground, the navy's work in this era underscored new interests that were not so much purely scientific or commercial as they were increasingly expansionist and tied to notions of growing national power. Voyages such as *Tuscarora*'s framed a Pacific empire that was increasingly defined by contests of power as much as it was grounded in naval science's traditional promotion of American commercial interests. *Tuscarora*'s voyage had briefly been delayed at San Diego in late 1873, for example, when news of the *Virginius* Affair over American-supported gunrunning to revolutionaries in Cuba nearly led to war with Spain. In February 1874, the ship called at Hawaii, where the Americans proposed to land the cable and construct a relay station. The islands had been the focus of American annexationist rhetoric stretching back to the antebellum era. While *Tuscarora* swung at anchor in Honolulu, Belknap watched closely as a monarchical election turned violent, and he landed *Tuscarora*'s bluejackets and Marines to keep order, to ensure the peaceful political succession of the Hawaiian monarch David Kalākaua, and of course to safeguard growing American commercial interests and economic investment in the islands.[42]

All of these were interconnected, rooted in the Navy's antebellum concerns for maritime commerce and foreign investment, but also indicative of new forms of American power at sea. To renewed calls for American annexation of the Hawaiian Islands in the 1870s, the American press declared, "the prospect that a Pacific-ocean cable would soon be laid was an additional reason in its favor." Cables formed a "web of power that tied the colonial empire together," wrote historian of technology Daniel R. Headrick, largely in the context of British imperialism. But the same could be said about the Americans' Pacific cable project. Such cables would be "an essential part of the new imperialism," Headrick continued, "they gave value to a handful of mostly deserted lands in the most isolated parts of the world." Commander Belknap himself added to the growing expansionist chorus. Later, he admitted, "I have been an annexationist since I saw the islands in 1874," referring to *Tuscarora*'s intervention there, and he believed that "Pearl Harbor can be made impregnable to naval attack." By the 1890s, he was writing to the *Hartford Times* in an undated letter, "if extending the borders and influence, the commercial and political power, of the United States is jingoism, what must we call the statesmen of the good old Democratic party who acquired the vast region of the Louisiana pur-

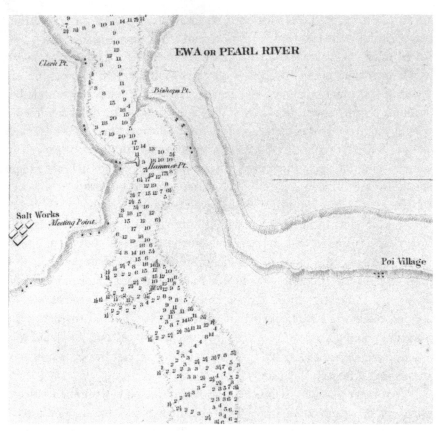

Detail of *Chart of Ewa or Pearl River*, United States Exploring Expedition, 1840, showing depth soundings and lines of triangulation in the channel, a village, and a meeting point, among other features indicative of broader commercial and cultural interests. Courtesy of the Library of Congress, Geography and Map Division.

chase, annexed Texas, invited war with Mexico, conquered and absorbed California, and concluded the treaty with Japan." As Belknap suggested, this was, in some ways, part of a continuous story of American expansion over the continent. While projects such as destroying the bar at Midway, surveying the isthmus, and sounding the sea floor were first and foremost commercial in nature, with roots in the navy's antebellum scientific work, they were also symptomatic of a new and growing American imperial power on the Pacific. "The spirit of Manifest Destiny went oceanic," write historians of science Michael Reidy, Gary Kroll, and Erik M. Conway of postwar American exploration at sea. In a way, it had always been oceanic, but Reidy, Kroll, and Conway are right to point to new and more power-

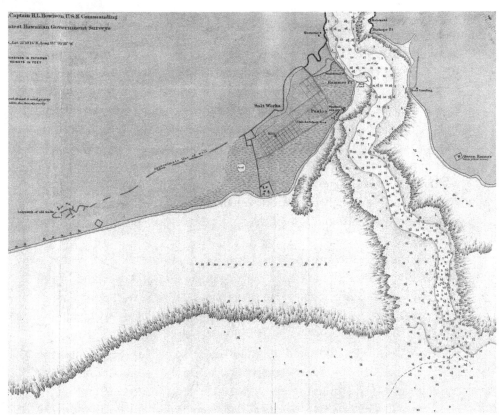

Detail of *Pearl Harbor and Lochs,* Hydrographic Office No. 1141, 1893, showing depth soundings in the channel and the process of transformation ashore, including "proximate line of wall," as Hawaii, Pearl Harbor in particular, gradually came to be a place of strategic interest for the navy and the United States. Courtesy of the Library of Congress, Geography and Map Division.

ful intentions bridging the mid-nineteenth century with the turn of the twentieth.[43]

The hydrographic chart itself reflected and perpetuated these changes. When Lieutenant Charles Wilkes and the U.S. Exploring Expedition surveyed the mouth of the Pearl River on the island of Oahu in 1840, for example, he wrote in the expedition's *Narrative* that "after passing this coral bar . . . the depth of water becomes ample for large ships, and the basin is sufficiently extensive to accommodate any number of vessels. If water upon the bar should be deepened," he concluded, "it would afford the best and most capacious harbor in the Pacific." Wilkes admitted that for the moment, though, the nearby port of Honolulu was "sufficient for all the present wants . . . and the trade that frequents them." There was "as yet . . . no necessity for such an operation," by which he meant deepening the shallow channel that led to the magnificent and nearly land-locked anchorage of Pearl Harbor. The expedition's chart, like its survey of the Fiji Islands, represented a mix of commercial and indigenous meanings. There was the bar itself and the soundings of Pearl Harbor's narrow channel. The Americans, however, had not extended their survey farther into the harbor. The bar blocked any passage and, as Wilkes said, Honolulu was sufficient for commercial purposes. The expedition's chart also marked a Meeting Point, a salt works, numerous villages, a block house, and oyster beds, evidence of commercial understandings that were so important to the early American presence in these islands, and also of the Anglo-American encounter with the indigenous Hawaiians.[44]

When the navy returned to Pearl Harbor again in 1887, the new chart— H.O. 1141—reflected the extraordinary changes that had occurred in the intervening half century. That year, King David Kalākaua—the same monarch whose claim to the Hawaiian throne Belknap and the *Tuscarora* had upheld in 1874—had signed a Treaty of Reciprocity with the United States. Among other things, it granted the Americans "exclusive right to enter the harbor of the Pearl River and establish a coaling and repair station for the use of vessels of the United States. And to that end," the agreement continued, "the U.S. may improve the entrance to said harbor and do all the other things needful to the purpose aforesaid." The steam sloop *Vandalia* began its survey soon after. "I had been interested in Pearl Harbor and the facilities which it possesses for the establishment of a secure and commodious coaling station," wrote Rear Admiral John G. Walker, chief of the Bureau of Navigation, to the secretary of the navy after ordering the survey. These, he thought, would be "of great value to the United States." The chart itself is markedly different from Wilkes's nearly a half century before. Gone are the Hawaiian villages and the oyster beds. Around the salt works, there appears a grid network of roads reminiscent of a modern town with only a broken and desultory "labyrinth of old walls" to mark the process of change on land. What happened to these effaced places and to the Hawaiians whose presence had once been implicit but so vital on the antebellum chart? Had they gone or were they simply obscured by the chart's new meanings? The focus clearly now lay on the water—on the coral bar, the circuitous channel, and the expansive body of protected water beyond, which appeared on *Vandalia*'s chart with a degree of comprehensiveness impossible in Wilkes's day. The implications seem clear. Here, the chart itself documents the process of change in this era as places like Hawaii became increasingly central to American aspirations in the Pacific.[45]

It would be another decade before the United States annexed Hawaii and yet another still before the Americans dredged the bar, deepened the channel, and constructed the beginnings of what would become a major American naval base at Pearl Harbor. These changes, however, which coalesced and took on greater momentum during and after the Spanish-American War of 1898 had their roots in the period after 1865. It was then that naval science, so central to the process of expanding American control over the sea, had begun to embrace a new, more muscular power rooted in the rhetoric of Senator William H. Seward and naval officers such as William Reynolds and George E. Belknap. From Alaska to the Arctic, Panama, Midway, and Hawaii—from the deep sea to the reefed littoral—this empire was framed by new understandings of the marine environment fixed by naval science on the hydrographic chart. Far from a period of stagnancy, this era was an expansive one for naval science, encompassing a multiplicity of pursuits that framed a Pacific Ocean in cartographic understandings that were commercial in nature, but also increasingly military and strategic, too. Of course, the marine environment sometimes humbled American efforts, but naval scientists' faith in scientific progress and the power of technology underscored their belief, to paraphrase *Tuscarora*'s Henry Cummings, that Old Ocean would, indeed, be conquered.

# 5 'Twixt the Devil and the Deep Blue Sea

. . . . . . . . . . . . . . . . . . . . . . . . . . . . . . . . . . . . . . . .

All nautical pride was cast aside as we ran our ship ashore
On the Caribee Isles, where the poo-poo smiles and the jumble gee
    chum chees roar.
We sat on the edge of a sandy ledge and shot the whistling bee-ee-ee
While the cinnamon bats wore water proof hats as they soused in the
    surf of the sea.

—"The Rollicking Window Blind," USS *Yosemite*, 1898

On February 5, 1901, Commander Joseph E. Craig and his command, the cruiser *Albany*, were in Hong Kong before a naval court of inquiry. The proceedings were to examine *Albany*'s grounding on December 17, 1900, as it delivered a detachment of Marines to the new naval station at Olongapo in Subic Bay on the Philippine island of Luzon. The sun had just set that evening over the mountains ringing the bay. The gathering darkness cast a uniform shadow across the water. Craig had consulted his chart, Hydrographic Office Number 1705. The gunner's mate called out distances from *Albany*'s new Barr and Stroud range finder. A lookout in the fighting top kept a sharp eye for discolored water ahead. Leadsman sounding in the chains reported no bottom at ten fathoms, plenty of water, in other words, to carry the cruiser safely ahead. Everything, according to Craig's report, seemed to check out. But at the moment he ordered a change of course for Olongapo, *Albany* shuddered slightly and stuck on an incorrectly charted shoal. The cruiser remained immobile for two days when, on the night of the nineteenth, after having discharged some two hundred tons of coal and fresh water and with high tide rising over the shoal, the ship finally floated free. Now Craig and his executive officer, Lieutenant Commander Albert G. Winterhalter, faced the court, hoping it would clear them of blame as they explained how *Albany* came to be laid up in dry dock while war consumed the Philippines.

For Craig, the strategically important waters of Subic Bay must have seemed a long way from the Hydrographic Office in Washington, but the two, in fact, were deeply intertwined. He could not have missed the irony.

Commanding the Hydrographic Office from 1897 to 1900, Craig had sat out the war revising old and often inaccurate charts, constructing new ones, and distributing these to the ships of the fleet. Frustrated, he longed for a wartime command. In January 1900, the Navy Department granted his wish: the thirty-four hundred ton-protected cruiser *Albany* with orders to fight the growing insurrection against American occupation of the Philippine Islands. But less than a year into his command, Craig was grounded. The former hydrographer of the navy blamed his chart.[1]

The chart and the marine environment would figure importantly in the coming proceedings and in a number of similar courts of inquiry, from Hong Kong to Key West, during the period between 1898 and 1901. The reports of naval commanders off Cuba, Puerto Rico, and the Philippines likewise testified to the failures of hydrography as the navy pressed wartime blockades into shallow coastal waters labyrinthine in navigational hazards. In nearly all aspects of naval operations in the Spanish-American-Philippine conflict— bombarding shore targets, engaging enemy gunboats, landing troops and supplies, cutting submarine telegraph cables, intercepting contraband, and simply steaming in and out of port—naval commanders were frustrated by the ocean environment and by the shortcomings of their charts. Seemingly small cartographic discrepancies of a mile or hundreds of yards were magnified when a foot or fathom might prove the difference between grounding and safe passage. Craig was hardly alone in blaming H.O. 1705, a chart that he himself had reviewed and issued as hydrographer. His court of inquiry, to which this chapter will return, is indicative of a conflict in which the natural environment was as dangerous to naval operations as the Spanish or Philippine enemy. The ocean environment proved a natural enemy, to borrow a term from the environmental historians of warfare Richard Tucker and Edmund Russell. Off the coasts of Cuba, Puerto Rico, and the Philippines, the navy had managed to achieve command of the sea, but only a tenuous command of the sea environment.[2]

In the last decades of the nineteenth century, the U.S. Navy experienced a revolution in strategic thinking and technological change that by the turn of the twentieth propelled the service into the top rank of the world's naval powers and coincided with the acquisition of a formal island empire that by 1899 stretched from Puerto Rico and Cuba to Hawaii, Guam, and the Philippines. Beginning in the 1880s, the navy had begun to replace wooden-hulled, sail-powered vessels equipped with auxiliary steam engines such as *Tuscarora* with new steel-hulled, steam-powered warships armed with breech-loading naval rifles. On the eve of the war with Spain in 1898, the fleet numbered six so-called coast defense battleships—Congress and the American people were still wary of anything that smacked too much of naval aggressiveness—seven cruisers, four monitors, an armored ram, sixteen torpedo boats, and six gunboats, which began to operate as a fleet rather than only in the traditional squadron system from which individual vessels had plied their time between foreign ports showing the flag and protecting American commercial interests. The creation of the so-called New Navy, a modernized fleet with still-limited but more offensive capabilities, reflected technological, intellectual, political, and cultural forces that coalesced at the end of the nineteenth century, culminating a process that, in some ways, began during the antebellum era, transformed after the Civil War , and became manifest in the 1890s, remaking the navy and the nation's vision of its place on the world stage.[3] In many ways, the turn of the century did mark a naval revolution in technology and strategic thinking, but one, nevertheless, with deep roots that were discernible, among other ways, in gradually changing ideas about the importance of the marine environment to American imperial interests. The American ocean empire, in other words, did not spring fully formed out of the war with Spain.

Central to this transformation was the publication in 1890 of *The Influence of Sea Power upon History, 1660–1783*, in which the naval officer and theorist Captain Alfred Thayer Mahan coined the term sea power and ultimately set off a frenzy of navalism in the navy, the halls of Congress, the White House, and in the governments and navies of nations such as Germany and Japan. With the possible exception of Harriet Beecher Stowe's *Uncle Tom's Cabin*, argued Mahan's biographer, "no book written in nineteenth century America by an American had greater immediate impact on the course and direction of the nation." Mahan, a lecturer and later president of the newly founded Naval War College in Newport, Rhode Island, a growing center of naval intellectualism and strategic thought, sought to distill the historical causes that propelled nations such as France and Great Britain to national power. Citing Britannia's rule of the waves, Mahan concluded that powerful nations, provided a certain geographical orientation to the sea, were often great sea powers. This he defined as having a vigorous maritime trade, a system of colonies and coaling stations, and a powerful navy. "It is not the taking of individual ships or convoys, be they few or many, that strikes down the money power of a nation," Mahan wrote, dismissing the navy's traditional strategy of *guerre de course*. "It is the possession of that overbearing power on the sea which drives the enemy's flag from it, or allows it to appear only as a fugitive; and which, by controlling the great common, closes the highways by which commerce moves to and from the enemy's shores." In war, sea power, Mahan contended, existed to close Maury's highways. Mahan's ideas were not new in 1890, but he was able to articulate them in a way that resonated within a particular political, economic, cultural, and technological moment ripe for the emergence of American navalism and imperialism.[4]

In sea power, Mahan had found a central concept that galvanized a political, military, and popular readership and framed the sea in new terms that had significant consequences for the ways the navy and the nation encountered the ocean environment both at sea and in the mind. The historiography on Mahan has focused mostly on his work as a theorist of naval strategy central to the emergence of American sea power and its attendant oceanic empire, a body of scholarship that, as one recent contributor argues, has generally misunderstood the nuance, complexity, and evolution of Mahan's ideas. Nevertheless, the lessons that readers drew from it in its time and later—however superficial—were powerful, condensed by Mahan in an introduction to *The Influence of Sea Power upon History* that identified various principles by which a nation could achieve national greatness through sea power. Often overlooked in theoretical discussions of Mahan's writing, however, is the cultural resonance of the notion of sea power itself, which not only convinced many Americans of the virtues of battleship fleets and island empires but also of the muscular notion that the sea itself could be controlled. Like Maury, Mahan was a master of conveying ideas using the written word, a skill that, in part, accounts for the credence that navalists increasingly put in his writings. He chose the term sea power, he later wrote, "to compel attention, and I hoped, to receive currency. I deliberately discarded the adjective, 'maritime,' " he admitted, "being too smooth to arrest men's attention or stick in their minds." Sea power, in other words, invoked dominance, seized the reader's attention, and elicited a response. Here we see Mahan the writer and conceptualizer, thinking about the ways in which his ideas could most effectively be conveyed. By adopting the concept of sea power, Mahan chose to frame his strategic theories in resonant ways and, in doing so, he subsequently and perhaps unintentionally recast the way the navy and naval science came to see the marine environment in this new era of American empire.[5]

In *The Influence of Sea Power upon History* and subsequent books and articles through the 1890s, Mahan militarized the sea, framing it in ways that in the abstract emphasized the ocean environment's military and strategic value and recast it in a new lexicon of dominance, control, and power that no other American naval officer or strategist had articulated with such force of words. Where Maury's common highway had been open to all, its winds and currents abetting commercial voyages from which American merchant and whaling fleets extended American economic and cultural influence over the world's oceans, Mahan's navy sought to close those highways to enemy commerce by framing the natural world as a space to be controlled in strategic terms. Mahan dubbed one of his principles of sea power "physical conformation," which included "natural productions and climate." These, he contended, were vital to a nation's sea power, but would also render that nation vulnerable in time of war. "Numerous and deep harbors are a source of strength and wealth," he wrote, "but by their very accessibility they become a source of weakness in war." The natural world, then, was "a physical condition" that "either gives birth and strength to sea power, or makes the country powerless." In Mahan's prose, the sea became a military "frontier." The Caribbean, he wrote, represented "the very domain of sea power." Cuba and the West Indies were "fortresses" guarding the entrances to an as-yet-unbuilt canal. In the Pacific, Mahan contended that Hawaii's value lay in its position as an "outpost of the canal." Mahan conjured a *mer militaire*, studded with fortresses, outposts, and frontiers providing support and defense for a revitalized commercial fleet and a powerful modern navy.[6]

To Mahan, notions of sea power as they related to the natural world remained mostly abstract, but to subsequent naval commanders whose operations in the Spanish-American-Philippine War sought to make his ideas manifest, knowledge of the sea environment would be central to achieving sea power in its real, watery dimensions. The bookish Mahan was more comfortable at his desk at Newport than at sea. He remained averse to the naval officer's rotational sea duty, a predilection that ran him afoul of some officers who still defined the profession's duties from the bridge of a warship rather than in the intellectual pursuits of science or strategic theorizing and authorship. Long before Mahan's assignment to the War College staff, he characterized his own commands at sea in his private correspondence with a dread and fear for the ocean's power and a keenly felt weakness of his own abilities as a navigator that, at times, bordered on incapacity. Perhaps it is not too far-fetched to conclude that when Mahan considered the marine environment in the course of his thinking and writing, he kept it at an abstract distance. He did have some limited hydrographic experience, leading a survey of Puget Sound in 1889 to determine a desirable location for a naval station there. Such work prompted his biographer Robert

Seager to quip that there was "nothing like a lead-line well swung . . . to sharpen an agile mind." While Mahan himself was not enamored with hydrographic surveying, we can also see in Seager's dismissiveness the ways hydrographic and scientific pursuits of any kind seemed at the margins not only of the naval officer's profession, but of the naval historian's as well. Hydrography seemed a rather meaningless aside to the more significant work of commanding ships and writing books that, with the benefit of historiographical hindsight, seemed to prophetically point the way to a coming century ultimately dominated by American naval power.[7]

In the Caribbean, naval operations were almost immediately beset with hydrographic difficulties. In January 1898, five ships of the North Atlantic Squadron, under the command of Rear Admiral Montgomery Sicard, the same officer who commanded USS *Saginaw* when it met the reef at Ocean Island in 1870, arrived in Caribbean waters for winter maneuvers and, ostensibly, to monitor the ongoing Cuban revolution against Spain. The problem of an adequate base of operations in the Caribbean plagued the navy from the start. French E. Chadwick, captain of the cruiser *New York*, borrowed a Mahanian term when he observed that the navy's station at Key West was "in no sense a stronghold," except, he added, "from the difficulty of navigation from the reef to the town." For Chadwick, Key West's physical environment of shallows and reefs might be regarded in militarized terms. For the defender, such a maze of natural obstacles could be a tactical or strategic asset, rendering navigation—to say nothing of combat operations—hazardous. But the same natural features that rendered a place easily defended, also made it undesirable to the navy, whose ships had to routinely steam in and out of port.[8]

Dry Tortugas, sixty miles to the West, offered a deeper anchorage, but had its own environmental challenges. On January 27, as the squadron steamed in Tortugas's Southeast Channel, the battleships *Texas* and *Iowa* ran aground. Two courts of inquiry into the groundings cited "the imperfection of the survey and chart." The court laid the chart's imperfections at the feet of the Coast and Geodetic Survey, which had surveyed the waters for the navy's use. These accusations grew into what would become by 1900 an open feud in which the navy claimed the survey's incompetence in charting waters of new strategic and imperial importance. Both ships' crews were absolved of blame in the kind of incident that, according to naval regulations, could be disastrous for the professional careers of naval officers. Francis J. Higginson, captain of the battleship *Massachusetts* and president of one court, immediately grasped the strategic consequences of the groundings in a letter to Assistant Secretary of the Navy Theodore Roosevelt. "The channels we were using were improperly surveyed and improperly buoyed," he informed Roosevelt. "I sincerely hope that while these vessels are in dock under repair, that no foreign complications will arise. Can not you stay the hand of war until we are prepared?" Higginson pleaded.[9]

More than the battleships, which suffered only minor damage, the groundings had apparently bruised the psyche of the squadron. After appealing to Roosevelt for new surveys, Higginson turned to the crux of the matter. "We all felt very blue the day of the accidents and it seemed as if it was not only raining but pouring. It created too a nervous distrust of the whole place and even now we are all shy of discolored water whether shoal or sunshine. We are taking no chances," he concluded. If Higginson spoke for his fellow captains, the North Atlantic Squadron seemed incapacitated by hydrographic uncertainty. Even with the channel buoyed, he remarked that the battleships should "enter and depart through it safely, I think," he added with palpable doubt. "We only know, or think we know, this one channel but 'there are others.'" Higginson then traced this growing paranoia farther up the chain of command to Rear Admiral Sicard himself. By February, Sicard was already an ill man. He had been suffering from malaria and had taken a short leave of absence from the fleet. But according to Higginson, the groundings of his battleships, which must certainly have reminded him of his sojourn as a young commanding officer on Ocean Island, were too much. "I think it was these accidents and the anxiety about more to come every time he moved his fleet which broke Sicard down," Higginson confided to Roosevelt. At the end of March, as war loomed in the wake of the *Maine* sinking and the De Lôme letter, Secretary of the Navy John D. Long promoted William T. Sampson, captain of the grounded *Iowa*, to rear admiral and command of the North Atlantic Fleet. Sicard was on his way to Washington and a seat on the Naval War Board, Long's strategic advisory council, apparently unfit to bear the stresses of command, environmental or otherwise.[10]

The Naval War Board, whose ad hoc membership at times included Sicard, Mahan, and Roosevelt, among others, advised Long to establish a close blockade of Cuba's north coast, an operation that required many shallow-draft ships and littoral steaming in ill-charted waters. To Mahan, blockade was essential to victory; it was the ideal manifestation of sea power. "Whatever the number of ships needed to watch those in an enemy's port," Mahan had written in 1895, "they are fewer by far than those that will be required to protect the scattered interests imperiled by the enemy's escape."

And so, while the navy did not yet know the whereabouts of the Spanish fleet, it set about establishing a blockade to strangle the Cuban garrison and force the Spanish to commit a naval force to the Caribbean.[11]

On April 22, three days before the official declaration of war, Long cabled Sampson at Key West to begin a blockade of Cuba from Bahia Honda in the west to Cardenas in the east, setting in motion naval operations in shallow, ill-charted waters. On the 23rd, the Americans appeared off the port of Cienfuegos on Cuba's southern coast. In all, Long and the Navy Department acquired more than one hundred vessels to augment the blockade. While the navy's battleships and cruisers steamed offshore, the inner blockade largely rested on this motley fleet of yachts, tugboats, revenue cutters, and ocean liners, hastily painted drab gray and armed with whatever guns the navy yards could bolt down. "About everything that could float and carry guns," one officer remarked, "was pressed into service." This was the fleet that gathered off Spanish ports on the north coast of Cuba. It has garnered little interest among historians of the navy and the Spanish-American War generally. A closer look at its operations, however, reveals the ways the natural world and the incomplete knowledge of it set down on the navy's hydrographic charts effected the course and consequences of naval operations in wartime.[12]

The Hydrographic Office supplied naval commanders and navigators with a stock of some thirty charts of the West Indies, but these largely reflected commercial understandings that had long informed the navy's hydrographic work and proved of limited use in wartime. Cuba's principle maritime ports—Havana, Santiago, and Cienfuegos—had been relatively well charted. But American hydrography of lesser ports and Cuba's long and hazardous coastline relied on old and inaccurate knowledge, derived, in many cases, from Spanish surveys. After the war, Captain Royal B. Bradford, Chief of the Bureau of Equipment, went so far as to wonder whether the Spanish had intentionally left Cuba's coast in such a state to deter an enemy fleet. For naval officers such as Bradford, who, as we will see in chapter 6, increasingly came to think about the environment in strategic terms, Spain's own hydrographic failures could be spun into some calculated plan to undermine American command of the seas around Cuba. Charts erred by miles. Among other things, the number of depth soundings remained scant in many places. On a postwar survey, one American officer found his Spanish chart "apparently made by running a few lines of soundings and then sketching in the coasts and cays by eye." Such inattention ran counter to the supposed precision of modern surveying method going back to the days of Wilkes's Ex. Ex., when mastery of the littoral—whether for trade or, now, for military conquest—rested in the cartographic ideals of accuracy and comprehensiveness. That the Spanish achieved neither suggested not just Spain's own cartographic failures but something of the weaknesses of its empire more broadly. Like the Americans' antebellum views of the Fijians whose own cultural practices reflected the dangers of the reefs offshore, Spain's inattention to the hydrography of its own empire seemed to preface the continued crumbling of Spanish colonialism. In this way, Cuba appeared ripe for the taking, an Old World empire on which the Americans would improve, in part, through more accurate surveys and navigational charts.[13]

The marine environment was vast and always changing, which belied the accuracy of both Spanish and American charts. Where sea met land, the hydrography was complex. Detailed surveys required months to complete and many more to construct and publish the finished chart. "No surveys have been recently carried on in any [country] and there are none now in progress," reported Commander Craig from the Hydrographic Office. He was referring to the United States and to British and Spanish surveys as well. The Hydrographic Office, then, could do little when commanders wrote requesting better charts "giving the coast of Cuba in more detail." One chart, a captain complained to Long, "lacks many of the details of the coast which would assist materially in inshore work." Others urged the Hydrographic Office to augment their stock with one or another privately made charts supposed to be "particularly good of the coast lines inside of the reefs." Short of this, however, most officers made due with the charts in hand, employed a Cuban pilot, or relied on an almost constant use of the sounding lead, a sharp eye, and a steady hand at the helm to navigate unknown waters.[14]

Given the enlarged fleet and an expanded theater of operations extending from Puerto Rico and Cuba to Guam and the Philippines, the Hydrographic Office could hardly keep pace with demand, to say nothing of revising and constructing charts. Craig reported "extraordinary demand" for the year. In three months of war, the Hydrographic Office distributed 43,910 charts, some seven times the normal peacetime number. "The energies of the office," Craig wrote, "were largely diverted from that part of the work . . . that results in the issuing of new publications of charts and of sailing directions covering new ground." Moreover, Craig responded to hundreds of requests from private citizens and from groups ranging from the Historical Society of Topeka, Kansas, to Johnson and Wood Hardware of Corsicana, Texas, requesting charts "to know something of the fleet movements." Above its naval importance, this emerging imperial cartography gave the American public its first sense of the war's geography and strategy, as armchair admirals breathlessly followed the movements of the fleet in the American press and turned to their charts of Cuba or the Philippines issued by the Hydrographic Office. Just as in the antebellum era, when the charts and texts of naval science reached a broader public audience through the work of American writers, in matters of turn-of-the-century naval warfare and empire building, the navy's charts helped the public envision the field of battle, the course of grave events, and, later, the expanse of American empire in the immediate aftermath of war. As I argue in the following chapter, the navy's charts, apart from their hydrographic or navigational value, took on broader, more abstract meanings that merged with ideas of Mahanian sea power and the militarization of the natural world.[15]

Naval commanders on the blockade, however, quickly understood that they could not trust their charts and had to rely on hunches and keen eyes as they wound their way through a maze of environmental hazards. Aboard the lighthouse tender *Maple*, executive officer Henry A. Wiley found trying moments of navigation more common than combat. As Wiley was discovering, blockade operations could be monotonous. *Maple* was little more than "a glorified tugboat," he admitted, and he often found himself in charge as his commanding officer, a lieutenant named Kellogg, apparently lounged the war away "in his underclothing, violently fanning himself with a palm-leaf fan." Among Kellogg's eccentricities, Wiley could never seem to wake him from sleep. One evening, with orders to steam in and out of Cardenas on Cuba's north coast, Wiley balked. "When I read these instructions I concluded that the captain was ill, that he was mentally fagged by nightfall and something in his mental make-up didn't function." The waters off Cardenas, Wiley knew, "contained no aids to navigation. It was shoal and full of shoaler spots called nigger heads," he remembered. "We could not possibly perform any useful service by going in there and might do irreparable damage." Unable to wake the captain, Wiley loosed the anchor in desperation as a storm began to blow. Awakened by the sudden jolt, Kellogg emerged from his cabin in time for Wiley to protest and save the vessel from possible grounding.[16]

For Wiley, navigating Cuban waters became more an act of the mind, relying more on feelings, premonition, and senses, than on the chart itself. Sometime later, *Maple* was again steaming in uncharted waters on the north coast of Cuba when it ran into shoals, and Wiley turned once more to the anchor. "We were sailing merrily along when I suddenly had a feeling that we were getting into shoal water," Wiley recalled. He likened it to "what a poker player would call a hunch," a sort of sixth sense familiar to commanders operating in ill-charted waters. Soundings reported ten feet. *Maple* drew nine. Wiley's hunch was prescient. "I let go the anchor," Wiley wrote. "Out came the captain." In Wiley's experience, the blockade proved a kind of "comic opera," played out by men who, without much else of pressing concern, battled natural foes more than the enemy.[17]

When commanders did engage the Spanish, however, the marine environment was less a comic opera than a serious hindrance. As Commander Chapman C. Todd looked from his chart to the water and then strained to see the Spanish gunboats anchored in the distance, he was aware of the hydrographic challenges that his small force faced in storming Cardenas Harbor. On the afternoon of May 11, *Wilmington*, Todd's third-rate gunboat, in company with the smaller revenue cutter *Hudson* and the torpedo boat *Winslow*, attacked Cardenas. They intended to destroy shipping in the harbor and to sink the gunboats. During the battle, heavy enemy fire damaged *Winslow* and killed five of its crew in one of the more intense naval actions of the war. Prior to the attack, however, Todd had no doubt consulted his copy of the Hydrographic Office's *Sailing Directions Caribbean Sea and Gulf of Mexico*. Like Wiley, Todd found the entrance to Cardenas a natural enemy that aided the Spanish defenses and left him in a precarious tactical situation. "The entrance to this bay," the directions read, "is so blocked up by small cays and shoals that it is only navigable for vessels of about 11 feet." *Wilmington* drew almost ten. "Even the most recent charts of this locality are not to be strictly depended on," the directions warned. And so, not trusting his chart, Todd was compelled to send the lighter draft *Hudson* and *Winslow* ahead to sweep for mines and to sound a channel through which *Wilmington*, with its larger guns, could pass.[18]

At Cardenas, safe passage depended not on the chart, but on the navigational pragmatism of American officers and a fortuitous boost from the marine environment. The two vessels set out on their hydrographic reconnaissance before noon on the eleventh. The Spanish had mined two of the three entrances to Cardenas, leaving the Americans a third "unexplored" channel. It was, according to Lieutenant John B. Bernadou in *Winslow*, "the shallowest of the three." The chart indicated one and three-quarters fathoms, or nearly twelve feet, at its shallowest. As Todd and Bernadou knew, however, the chart could not be trusted. With high tide in the Americans' favor, they could perhaps count on an additional one and a half feet to get *Wilmington* through. "If this depth of water actually existed, and if the soundings shown upon our chart were correct, then entrance through this passage for vessels of *Wilmington's* draft was safe and practical at high tide," wrote Bernadou. He was hopeful, but not altogether confident, that the chart and the tide would work in the Americans' favor. *Winslow* and *Hudson* steamed slowly through the channel, accompanied by a Cuban pilot with knowledge of the local waters while soundings were "constantly taken with the lead." Bernadou had found a channel of ten feet, just enough water with the tide in the Americans' favor to carry *Wilmington*. But as the vessels turned to sweep the channel for mines and report their survey to Todd, *Hudson* grounded and "hung" for some time on a shoal. Shifting weight and adjusting trim, its crew was able to get the vessel afloat again, but Bernadou reported that the minesweeping "could not be done on account of the grounding."[19]

*Wilmington* thus entered Cardenas Harbor guaranteed little more than inches under its keel and the possibility of mines ahead, to say nothing of the threat from the Spanish gunboats once safely within the harbor. Running the channel proved predictably harrowing. High tide came just after noon, and the three vessels steamed for the channel—*Hudson* and *Winslow* on *Wilmington's* starboard and port bows, "to give warning in the event of the discovery of any sudden shoaling of the water." All eyes were on *Wilmington*, which stopped, started, and stopped, then started again, proceeding slowly and cautiously over the shoals. "The stirring up of the coral mud and the resultant whitening of the sea," Bernadou observed from *Winslow's* bridge, indicated "that there was very little water left beneath her keel." But the Americans emerged unscathed for the attack, and the enemy had failed to fire on them as they weaved slowly through the hazards. Had the Spanish seized that opportunity, or mined an already dangerous passage, the Americans perhaps would not have succeeded. As in so many other aspects of the naval war, American success depended on Spanish negligence. To judge from Bernadou's report of the action to the secretary of the navy and a subsequent article he penned for *Century Illustrated Magazine*, one important enemy the Americans faced was a nonhuman one. A reading of these events through Bernadou's eyes—for that is all he had to navigate the ill-charted passage—suggests an action fraught with environmental peril that, along with a more formidable enemy capable of exploiting the nature of the marine environment and the Americans' ignorance of it, might have proved decisive for Spain at least in the waters off Cardenas.[20]

At the same time, American vessels were scouring the waters off Cienfuegos in search of submarine telegraph cables, an operation that added a new underwater dimension to the blockade and the larger strategy of the naval war. The cables off Cienfuegos linked Cuba's southern coast by way of Havana with Jamaica and Madrid. But the cables lay on the sea floor, requiring a deep knowledge that the Americans lacked. "Outside of the records of the proprietary cable company and excepting as to some shore ends, the precise location of every cable is unknown," noted Captain Caspar F. Goodrich, who spent much of the war cutting underwater cables around Cuba in the auxiliary cruiser *St. Louis*. "No chart that I was able to obtain, no source of intelligence," he continued, "could tell me the very spot to go to for the purpose of raising the submarine wires I wished to sever." In qualifying the chart as a source of intelligence, Goodrich captured the transforming role of hydrography in the era of the New Navy and the new American empire. While the chart and the knowledge implicit in it had long been a vital strategic and tactical asset in war, by the end of the century, charts formed part of a larger, emergent culture of naval intelligence within the Navy's institutional structure that formalized the acquisition of knowledge as a strategic asset. The Office of Naval Intelligence, for example, had been created in 1882 to begin to collect all manner of information vital to growing American strategic and national interests around the world. By the end of the century, charts of strategically important waters— in contrast to the traditional commercial imperatives that long informed hydrographic surveying and charting—underscored the larger transformation of the navy, naval science, and the new empire it sought to defend.[21]

Off Cienfuegos on May 11, as Spanish bullets hissed in the water around him, Lieutenant Cameron M. Winslow must have cursed his chart. Charged with commanding boats from the cruisers *Nashville* and *Marblehead*, Winslow pressed his small command into shallow waters a few hundred yards from Spanish batteries ashore. "Keeping a good lookout for rocks and reefs, the boats pulled steadily on," Winslow wrote, "the inaccurate Cuban charts giving us little information as to the distance from the land at which we should find shoal water." At a depth of twenty feet, Winslow's men spotted the cables they sought to sever. The Americans dragged, grappled, cut, and pulled the cumbersome iron cables out to sea where they could not be retrieved while Spanish guns took easy aim from shore. Winslow was wounded in the hand. Without accurate knowledge of the sea floor, the Americans had little choice but to expose themselves to fire in shallow waters where they could physically see the bottom. "To cut the enemy's lines of communication is always important," Winslow concluded. It was an old military axiom, but one given a new, more challenging underwater dimension off Cuba that begged more detailed knowledge of the sea floor itself.[22]

The sea floor on which these cables rested became one more marine environment newly relevant to naval war. After the war, Rear Admiral Charles H. Stockton, a professor of international law at and later president of the Naval War College, immediately began studying the legal standing of these cables and the grounds on which they could be destroyed. The issue concerned not only private ownership, territorial waters, and freedom of the seas but also the extent to which the ocean floor itself was subject to national control. Stockton pointed out that during the Franco-Prussian War, 1870–71, and the War of the Pacific, 1879–83, the belligerents cut cables "both within territorial waters and in extra-territorial waters or the high seas." The former struck Stockton with "little doubt" as "the right of the belligerent." But the issue of national cables in international waters raised other issues. Indeed, submarine telegraphy begged the question of who controlled the deep sea floor not just in the abstract, imagined ways that so informed *Tuscarora's* deep sea sounding voyage in the 1870s, but now from a legal and political perspective. As long as these depths were largely inaccessible to all but the most advanced methods of surveying and science, the bottom remained largely immune to war making and questions of political or legal control. But as new technologies permitted scientists and hydrographers to encounter the sea floor and brace it with cables that were important during peacetime but also strategically vital in war, this most remote of earthly environments increasingly became opened to questions of control and conflict.[23]

While the notion of sea power itself did not explicitly frame the sea in territorial terms, it nevertheless suggested that national power could be extended across the sea and, during and after the war, even to the marine environment itself. Submarine telegraphy extended sea power to the sea floor. In 1900, a captain in the Army Signal Corps published an article in *Naval Institute Proceedings* with the Mahanian-sounding title "The Influence of Submarine Telegraph Cables upon Military and Naval Supremacy." In it, he referred to "a great sea division" having "no better guides to boundaries than the submarine cable networks." As cables framed the sea floor, spanning vast underwater spaces and connecting far-flung empires, they represented new claims to the marine environment. National and military power could pierce the water's surface and extend to the depths below. The deep sea and the sea floor thus entered the navy's strategic discourse, becoming one more dimension of naval warfare, cartographic control, and imperial domain.[24]

The limits of hydrographic knowledge, however, hampered American command of the sea from Cienfuegos all along the southern coast of Cuba, a stretch of water notorious for shoals that afforded havens for blockade-runners and frustrated the Americans pursuing them. "The natural conditions existing in these localities," wrote one officer, offered "great advantages" to the enemy. On June 28, President William McKinley declared an extension of the blockade on the southern coast of Cuba from Cape Frances in the west to Cape Cruz in the east, a stretch of five hundred miles, encompassing two-thirds of the southern coastline. Particularly challenging to navigation was a gulf of shallows and shoals sheltered by the Isle of Pines and flanked by Cape Frances and the port of Batabano with a rail connection northward to Havana. The channels through this gulf carried no more than "12 or 13 feet of water" according to the navy's sailing directions. It warned of "almost innumerable cays and sand banks, as yet very imperfectly known, and forming intricate and numerous channels. To navigate these channels and to identify the cays used as landmarks," the directions concluded, "local knowledge is positively necessary."[25]

The yacht *Eagle* pressed the blockade into this environment on July 12 when it chased the Spanish steamer *Santo Domingo* aground on a shoal and, having little choice, destroyed it to prevent its recapture. In the late morning, Lieutenant William Henry Hudson Southerland, commanding *Eagle*, spotted the sleek black steamer on the horizon between Cape Frances and the Isle of Pines, running northward toward Batabano at high speed. *Eagle* gave chase, and *Santo Domingo* grounded, unable to outrun the Americans while navigating the shoals. "With an uneven coral bottom of varying depth, and, with boats sounding ahead," Southerland reported, the *Eagle* "made slow progress until within about 2,000 yards of the steamer." *Santo Domingo*'s crew had already abandoned the ship by the time the *Eagle*'s whaleboat pulled alongside. The Americans discovered it to be laden with "munitions of war" and an "immense amount of food supplies." With the steamer hard and fast on the reef, the crew doused the ship with kerosene, opened its magazine, and left the *Santo Domingo* an inferno that smoldered for weeks.[26]

In deciding to destroy the vessel, Southerland was governed by the dangers of the surrounding waters and the limits of his hydrographic knowledge. As he noted, the *Santo Domingo* might have been a considerable prize, which, under the rules still governing the capture of these vessels, would have brought a fortune divided among the Americans. "I do not think I am far wrong in stating that if the vessel and cargo could have been saved and brought into port the appraisal value . . . would have fallen but a little short of $1,000,000," Southerland reported. It would have been the kind of payoff that made the drudgery of blockade duty bearable. It was certainly with regret that Southerland cited the hydrographic concerns, which influenced his decision to destroy the prize. He first pointed to the *Eagle*'s inability to pull the much larger vessel off the reef and then cited "the possibility of an attempt at recapture, which I think *Eagle* could have resisted had it been possible to maneuver a 12-foot [draft] vessel on those unknown coral shoals at night with rapidity and safety." Southerland rightly placed little faith in his chart, and even less in the ability of his vessel to reclaim the Spanish vessel given the environmental challenges he found himself in. The natural environment had forced Southerland's hand, denying the navy an important capture and the *Eagle*'s crew the spoils of war.[27]

The Americans were similarly thwarted in their attempts to provide supplies to Cuban insurgents on land whose attacks against the Spanish army were the only force ashore until the U.S. Army arrived in late June. On the evening of May 30, not far from where Southerland would destroy the Spanish steamer in July, Lieutenant Commander Daniel Delehanty and the crew of the yacht *Suwanee* were preparing to land supplies for the Cuban army from the paddle-wheel transport *Gussie*. Under cover of darkness, Delehanty steamed slowly inside the cays, "there being no reliable chart of this part of the Cuban coast." The reported presence of Spanish gunboats nearby added more danger to an already tense situation. But Delehanty had reason to think he might succeed. He had two Cuban pilots aboard—one, in Delehanty's words, who professed to be "a very competent pilot in the waters in which we were to operate." Lacking an accurate chart, such pilots were supposed to be an essential source of local knowledge about coastal waters. *Suwanee* grounded nonetheless, and Delehanty could not free the yacht until high tide finally rescued the hapless vessel the next morning, having spent the better part of the night immobile, an easy target had it been discovered and attacked by Spanish gunboats. In the morning, Delehanty concluded that *Gussie*, with its larger draft, would no doubt meet the same dangers. He therefore called off the operation, citing "the impossibility of landing the supplies with the means at our disposal."[28]

By August 12, the navy had strengthened the blockade on the southern coast sufficiently to attempt an attack on the stronghold of Manzanillo, a port in southeastern Cuba, but once again the effort demonstrated the limits of the Americans' hydrographic knowledge and the pragmatism needed to overcome it. The navy had already attempted two raids on the port on June 30–July 1 and July 17. These had done moderate damage, but ultimately demonstrated the strength of the Spanish gunboats and the city's garrison and defenses. After the July 1 raid, one officer had written Secretary Long of his "regret that we could not steam right past the city and endeavor to sink the gunboats as we went along. But we knew nothing about the channels and had to return by the one we had found by the use of the lead and the appearance of the water." This time, the navy intended to take the city and perhaps force an end to the war, which had not come as expected following the defeat of the Spanish fleet on July 3 at the Battle of Santiago de Cuba and the surrender of Santiago to the army on July 17. On August 8, a force of six vessels under the command of Captain Goodrich in the cruiser *Newark* assembled at Cape Cruz and prepared to take Manzanillo. Again, the Americans met environmental challenges that they could not ultimately overcome. "It was . . . evident that to take a vessel the size of the *Newark* within bombarding distance of the town was an undertaking beset with danger and possible disaster," observed Lieutenant William F. Halsey Sr., *Newark*'s executive officer and navigator. Nonetheless, he continued, "risks were to be taken that in time of peace might be deemed inexcusable . . . war conditions demanded them, provided the necessary nerve and ability were combined." Here, Halsey was referring not to the threat of Spanish gunboats or land batteries, but rather to the hazards of the marine environment. All this bravado about danger and disaster, nerve and ability, Halsey thought, would be tested most among the channels, shoals, and reefs, which together constituted Manzanillo's first—and perhaps its best—line of defense.[29]

The hydrography of the approaches to Manzanillo resembled so many Cuban harbors—a maze of sand, coral, and water for which the chart could not account. The bottom was irregular, Halsey wrote, "and of the currents no man is able to tell." A barrier of keys ran northwest from Cape Cruz, masking Manzanillo behind an inland sea of shoals called Buena Esperanza— "good hope," of course, was what the Americans needed in these waters. "To those not possessing a local knowledge," Halsey observed, "the keys in this vicinity have a strange similarity in appearance, and as the chart failed to show some that existed, and depicted others that neglected to appear, the difficulties of determining positions by bearings can be realized." Halsey was suspicious of the chart's "strange variance in soundings" and spoke for many fellow officers when he expressed "doubts upon the reliability of this important aid in navigating." The Balandras Channel was the most direct route through Buena Esperanza, but it was only eighteen feet deep, and *Newark*, Halsey noted, drew twenty-two feet and three inches of water. Cuatro Reales Channel was the only alternative, but his

Detail of *Cuba and West Indies*, 1904, showing the Great Bank of Buena Esperanza
and the approaches to the port of Manzanillo, which the Americans attempted to
take in August 1898 without the aid of this chart constructed by the navy after the
war. The chart illustrates the complexity of navigation amid numerous shoals on
Cuba's southern coast. Courtesy of the Library of Congress, Geography and Map
Division.

sailing directions deemed it impassable—a sign, Halsey remarked, that
"was not reassuring." But a Cuban pilot had promised five and a half fathoms, or just over thirty feet. Steaming slowly with sounding leads drop-
ping on both sides of the ships, the Americans made a safe, if painstaking,
passage through Buena Esperanza, leaving pickle kegs anchored at sharp
turns in the channel to serve as makeshift buoys for the return passage.[30]

The ships sounded general quarters the next afternoon and commenced
the attack on Manzanillo, but the larger *Newark* was hemmed on all sides
by shoals and was forced to let the small gunboats maneuver among the

shallows. The channel rapidly narrowed as *Newark* opened fire with its six-inch gun. A report from the leadsman of five fathoms was too much for the Cuban pilot, who walked to the end of the bridge, Halsey recalled, "indicating that he washed his hands of all further responsibility." Now without a pilot and, for all intents and purposes, without a chart as well, *Newark* pressed on to four and a half fathoms when Captain Goodrich evidently lost his nerve. He ordered the engines reversed, but *Newark* stubbornly continued its forward movement. "The propellers did not have the full effect with the scant water," Halsey wrote. Goodrich let go the anchors until the ship finally stopped and then backed away from the danger. *Newark* could do little but support the smaller vessels as they pressed their attack, laying-to in five fathoms until daybreak, August 13, when the beleaguered Spanish brought word of the armistice and the end of the war. The attack on Manzanillo had demonstrated the concerted power of the blockade, but throughout the entire operation, the Americans were slowed by shallows and plagued by uncertainty brought on by charts they did not trust.[31]

The navy encountered similar problems off Puerto Rico along coasts as poorly charted as those navigated by Goodrich and Halsey at Manzanillo. On May 12, Rear Admiral William T. Sampson, in command of American naval forces in the Caribbean, led a bombardment of San Juan with his battleships and cruisers. Even in this bustling maritime harbor, the admiral remained cautious. "The soundings laid down on the chart of the island were . . . doubtful," Sampson wrote, "rendering a near approach to the coast dangerous, except while in the usual track for entering or leaving the port." Worried about the chart and surely reminded of the groundings of *Texas* and his own battleship, *Iowa*, in January, which had apparently done in his predecessor Sicard, Sampson drew up a plan of battle that would eliminate as much as possible any hydrographic uncertainties left by his charts. He ordered the tug *Wompatuck* to steam ahead, leaving a flagged boat to indicate the point at which the Americans were to execute a change of course across the harbor. With the light draft cruiser *Detroit* leading the column of battleships and cruisers to indicate shoal water, the Americans steamed off San Juan, firing away at the city's defenses as men on each ship continuously sounded with lead lines on the unengaged side. None of Sampson's ships grounded in the bombardment, but the juxtaposition of battle and sounding, simultaneously executed on either side of these vessels, suggests how closely these two activities were associated during the Spanish-American War. To heave the lead with one hand while firing at the Spanish with the other was commonplace.[32]

The navy returned to Puerto Rico at the end of July as part of Major General Nelson A. Miles's invasion of the island, but it was uneasy about providing support for the army in uncharted waters. The original plan was to land troops at Cape San Juan on the northeast coast of Puerto Rico, but Miles changed his mind mid-voyage. Guanica, on the southern coast, he told Captain Francis J. Higginson, would catch the Spanish by surprise, and so he wished to be landed there. Higginson, who before the war had confessed to a fear of shoal water in his letter to Theodore Roosevelt, now faced the prospect of his battleship, *Massachusetts*, running aground off Puerto Rico to fulfill the whims of a general. He protested. "From a naval perspective," he told Secretary Long, "I could not so effectually cover his landing or protect his base at Guanica as I could do at San Juan." Higginson cited a number of environmental threats, none of which directly involved the Spanish. He referenced the *Massachusetts*'s deep draft and exposure to storms, and finally noted, "the south coast of Porto Rico [*sic*] was imperfectly surveyed, and lined with reefs." Miles, the commanding general of the army, brushed off Higginson's protests and the landing at Guanica proceeded without harm to the navy's ships. But Higginson reported to Long that *Massachusetts* had come uncomfortably close to grounding in nineteen feet of water where the chart had indicated a clear sixty-six feet. Higginson, of course, was correct to second-guess his chart, and *Massachusetts*'s near-grounding off Guanica only reaffirmed his continuing fear of lurking shoals. Echoing his plea in January at Dry Tortugas, Higginson once again requested surveys to be made, appealing this time to Secretary Long about the fleet's hydrographic troubles. "If the operations on the coast of Porto Rico [*sic*] are to be continued I would recommend that two surveying vessels be sent to that island as soon as possible," Higginson reported. The Spanish-American War ended on August 12 before the army could march on San Juan and before the Navy Department could act on Higginson's request. In once more citing the need for hydrographic surveys, Higginson's war had ended much as it had begun. The navy's charts were poor, and the fleet's experience during the war raised hydrography to new strategic importance, convincing many within the navy that one of the most important postwar activities, in seeking to defend the coasts of a new American empire, was to chart them.[33]

Meanwhile, war continued in the Philippines, which the Americans had seized at the beginning of the war when Commodore George Dewey's Asiatic Squadron steamed into Manila Bay and defeated the Spanish fleet there. Now, however, a Filipino-led insurrection against American rule pressed the navy again into dangerous coastal waters. "The Philippine Islands were not well surveyed, and it is unknown dangers which are most feared," observed Captain Royal B. Bradford, chief of the Bureau of Equipment, who supervised the Hydrographic Office. Military historian Brian M. Linn has observed that the operations of the navy's shallow-draft gunboats during the Philippine-American War was "exhilarating work," in which naval officers were busily "keeping the cranky engines working, securing sufficient coal, avoiding shoals and reefs, and stalking smugglers and pirates." Yet in avoiding shoals and reefs, Linn overlooks the significant hydrographic challenges the Americans faced and the relationship among the marine environment, cartographic knowledge, and American attempts to establish sea power throughout its new territorial empire.[34]

In fact, it was an uncharted rock that inflicted the navy's largest material loss of the entire Spanish-American-Philippine conflict. On November 2, 1899, the cruiser *Charleston* was steaming off the north coast of Luzon when it struck the danger, broke apart, and sank. All survived, but the ship was a total loss. The environment claimed what Spanish and Filipino arms could not. *Charleston*, ironically, had been returning from survey duty of its own in the archipelago—part of a broader, desperate effort to gain environmental knowledge in the middle of the growing conflict. "We were one of the first ships of our Navy to cruise along this coast," wrote a chief electrician aboard the cruiser. He then went on to reference the inaccuracy of the charts, which, he thought, "were years old and very unreliable. Some of the points were supposed to be taken from the Spaniards," he thought, "and were over a hundred years old." In December, a court of inquiry met to investigate the circumstances surrounding the accident. Not surprisingly, the proceedings focused on the chart, a British Admiralty publication dating to 1867. The best hydrographic information for the north coast of Luzon, in other words, was more than thirty years old. The court concluded that Captain Charles W. Pigman, *Charleston*'s commanding officer, had done everything by the book. He had consulted the chart and sailing directions, corrected the ship's bearings for magnetic variation, and sounded for uncharted shoals. No one and nothing were to blame, the court concluded, but the chart itself. Large vessels such as *Charleston* were not ideal for this kind of work, and, as Linn shows, the navy soon turned over the bulk of its operations in the Philippine War to a growing fleet of shallow-draft gunboats. *Charleston*'s loss, however, underscored the very real danger the marine environment represented, particularly to the larger vessels of the fleet, the sort that had given Higginson so much pause at Dry Tortugas and off Puerto

Rico. As the navy planned for future wars in Asia and the Caribbean, the ill-charted waters of the American empire seemed likely flash points.[35]

It was in the context of hydrography's wartime failures and its newfound strategic importance, then, that Commander Craig and Lieutenant Commander Winterhalter stood before the court of inquiry, examining Albany's grounding in Subic Bay. It was now February 1901. The Philippine War raged. Albany lay dry-docked in Hong Kong, its hull plates crumpled and disfigured amidships—a testament to operations in waters rife with cartographic errors. The court called three principle witnesses, a naval cadet who was officer of the deck at the time of the grounding, then Craig, and finally, Winterhalter. The question concerning the court was whether Craig and Winterhalter had erred in navigating Albany into the port of Olongapo or whether the chart's inaccuracies cleared them of any blame in the grounding.

After recounting the events of the previous December, the court moved to consider the act of navigation in detail, revealing just how deeply naval officers engaged the marine environment in the course of their duty. Steaming into port was among the most dangerous acts of seamanship, a moment in the voyage that required particular vigilance as the vessel moved into the shallows at the convergence of land and sea environments. In their testimony, Craig and Winterhalter recounted their preparations for Albany's approach to Olongapo. As navigator, Winterhalter was deeply aware of the natural world. He spoke of the declination of the sun, the phase of the moon, and the resulting effect these had on the tide. He noted the depth of water almost continuously and, once grounded, he inquired about the character of the bottom. He did all this while taking bearings on and distances from islands and other landmarks. Even in this era of steam and steel warships, navigation was a process fundamentally rooted in environmental knowledge. To the judge advocate general's question about whether there was any indication of shoal water, Winterhalter replied, "No, from the appearance of the water, I was sure there was none. I was looking for it," he assured the court, "that is a sort of instinct." Winterhalter had to account for an extraordinary number of natural factors, which were always changing on their own account and as the ship moved through the water.[36]

All this came together on H.O. 1705—the chart in question—which Craig and Winterhalter blamed for the incident. "The grounding," Craig testified, "was due to the incorrect charting of the shoal off Mayanga Island running as it does at least 150 yards farther out than shown on H.O. Chart 1705, the one used in navigating." Winterhalter echoed his commanding officer. The shoal, he said, was incorrect "certainly in extent, and very probably also in contour." The court then watched as Winterhalter, the chart spread before him, measured the distance of the shoal from Mayanga Island. He then did the same from the island to the spot where *Albany* had struck and estimated an error of three hundred yards between the shoal as it existed in nature and the shoal as it appeared on the chart. Winterhalter's measurements were the climax of the proceedings. *Albany*'s grounding seemed to turn on this moment—a cartographic and juridical spectacle on which two officers' careers hung, on which *Albany*'s hull had crumpled, and, perhaps more broadly, on which Americans' intentions of sea power, in part, rested. At the center of all this, of course, was the much-maligned chart H.O. 1705, which represented so many others, from Puerto Rico to Luzon, whose coastlines, depth soundings, and tidal information were incomplete, erroneous, or misleading.[37]

The officers recounted their misplaced faith in the chart. Asked by his counsel whether he felt confident in its accuracy based on the frequency with which vessels came in and out of Olongapo, Craig nodded. "I believed the charts of Manila Bay and of Subig Bay and the coast between these bays were an exception, in point of accuracy, to charts showing other portions of the waters of the Philippine Islands." Perhaps Captain Pigman, *Charleston*'s former commanding officer and president of the court, nodded in agreement. "All this," Winterhalter added, "after the harbor had been visited with frequency by our ships . . . from the time the ships of Dewey's squadron first explored the bay in search of the Spaniards to the day the *Albany* grounded, and not a note or correction to show that this much traversed spot had a shoal nearly if not entirely twice as great as that shown on the chart." Asked, "did you know where the ship was when she grounded," Winterhalter replied tellingly, "yes, I knew exactly where she was *on the chart*" [italics added]. There was a crucial difference, of course, between the sea and the sea as H.O. 1705 represented it. Winterhalter and Craig had placed their faith almost wholly in a chart that they assumed to be an accurate representation of Philippine waters. This was where they had erred.[38]

When the court presented its conclusion on February 11, 1901, after six days of testimony and deliberation, it reaffirmed what many naval commanders already knew and what Craig should well have grasped himself. H.O. 1705 was not to be trusted. This was the crux of the matter. "The cause of the grounding should be attributed to the fact that the shoal off Mayanga Island has extended to the southward and eastward probably more than 150 yards since the survey was made from which H.O. Chart No. 1705 was constructed," the court concluded. Winterhalter had set a course for *Albany* confirmed by Craig that would "allow for less than 250 yards as a margin of safety." This, "we deem . . . a grave error of judgment amounting to a fault," the court concluded. In other words, the court expected Craig, former hydrographer of the navy, to have placed so little faith in the charts of his own office as to assume that they were at least 250 yards in error. He should have set *Albany*'s course according to what the chart did not indicate rather than by what it did. A conscientious navigator, it seemed, would assume error to be as much a part of the hydrographic chart as depth soundings themselves. For the Americans, whose cartographic method since the days of Charles Wilkes had been predicated on precision and comprehensiveness, the Spanish-American War proved a watershed moment, exposing the navy's charts as inaccurate, incomplete, and—it seems—incapacitating for many naval commanders forced with tightening the blockade around Cuba or conducting operations off the coasts of Puerto Rico and the Philippines. Of course, H.O. 1705, like the chart that obscured *Charleston*'s ill-fated encounter with a rock, and so many others spanning the waters of the new empire, was not primarily American in origin. These charts were compilations of hydrographic data accumulated, in some cases, over centuries and by many different nations. Nevertheless, the war had shown that the Americans could no longer afford to dismiss such inaccuracies—as Bradford and others had—as the work of an incompetent enemy, or one who used inaccurate environmental knowledge as a strategic defense against attack. The Americans would now need to chart these waters themselves.[39]

During the Spanish-American War and its aftermath, hydrography and the marine environment presented challenges to naval operations that were different but no less dangerous than the enemy. Indeed, we might consider the marine environment, which figures so centrally here, to be a natural enemy whose dynamism and complexity influenced the process and sometimes the outcome of events. As environmental historians Richard Tucker and Edmund Russell argue, "the very usefulness of nature to one side of a conflict has often made it the enemy of another." At war's end, Rear Admiral Sampson praised the fleet, citing "the admirable navigation of the vessels under unfavorable conditions. They surrounded an island," he concluded, "the harbors and coasts of which were not well surveyed." In 1898, Craig himself had admitted as much. The navy's charts, he reported, were "lacking in details owing to untrustworthy data, and in some features they are no doubt erroneous and the channels amongst the keys for light draft vessels are known only to local pilots." Perhaps he should have heeded his own warning. Yet even as the navy expanded its surveys to encompass the waters of the new American empire, the sea remained a difficult place to chart and to navigate. Its vastness, depth, and complexity precluded the quick and immediate construction of charts for these waters. Its dynamism, its ebb and flow, and its ability to create and recreate coastal environments always rendered representation on the page elusive. Ultimately, hydrography could only partly capture an environment in constant motion. The chart represented both the navy's intent to command the sea and, ultimately, its inability to fully grasp this most basic imperative of sea power.[40]

The Spanish-American-Philippine War was not the first time American warships were drawn into shallows that were hazardous in their environmental dangers, but it did mark an important moment in which larger political, technological, intellectual, and strategic factors altered the relationship between naval science, the marine environment, and American empire. As the commercial imperatives that had long characterized naval science and exploration were subsumed by new strategic questions, a more offensively oriented navy, and, ultimately, the imperatives of defending a vast oceanic empire, the chart and naval hydrography assumed new strategic importance that increasingly excluded civilian science and theoretical questions with seemingly little direct bearing on the important work of empire building. As the Spanish-American-Philippine War ushered new roles for naval science, it also marked the culmination of a century of an uneasy civilian–navy partnership in ocean science that would not reemerge until after 1918. When the last gun fell silent over Spain's shattered empire, the United States and its navy emerged as a new world power. The Americans secured that empire, in part, through a Mahanian vision that began to frame the sea environment as a place to be controlled by battle fleets and whose coastal waters, at least, the United States could claim as American territory. Such visions did not initially dive deeply into the hydrography of sea power, but to those naval commanders charged with putting Mahan's ideas into practice, the marine environment and the hydrographic chart presented very real challenges to "controlling the great common." For the navy, the Spanish-American-Philippine War proved a heady moment with decisive victories that seemed to affirm the changes then coursing through it. Yet, if the Navy had been victorious, it nevertheless had to reckon with waters largely outside its control. The familiar course of the naval war, bookended by the battles of Manila Bay and Santiago de Cuba, rightly privileges major fleet engagements in relatively well-charted or deep offshore waters against weak foes. The often-overlooked experience of the inshore blockaders, however, adds complexity to this familiar narrative, suggesting the centrality of the sea and knowledge of the ocean environment to American sea power.[41]

In 1903, Captain Bradford, bureau chief and member of the newly formed General Board of the Navy—a sort of naval general staff—recalled the navy's role in the war. With a nod to his supervisory duties over the Hydrographic Office, he cited the environmental difficulties of the conflict. The war, he thought, "would have assumed a totally different aspect" had the fleet "possessed the charts which have been constructed from surveys made since the Spanish war." Bradford's was hardly a ringing endorsement of what historians have come to consider, with few qualifications, a decisive show of American naval force. Bradford's statement hints at "different aspects" of this conflict, such as the marine environment, that can shed new light on the way historians understand the growth of American power at the dawn of the twentieth century. Indeed, just as Bradford was writing his postwar assessment, the naval historian Edgar Stanton Maclay published his *History of the United States Navy, 1775–1902*, which he had revised to encompass naval operations in the recent war. In this book, which the Naval Academy soon adopted for use by its midshipmen, Maclay cited "perilous shoals" and charts "so unreliable as to be worse than useless. Well might it have been said," he surmised, "that our officers, seamen and ships engaged on this service were placed ' 'twixt the devil and the deep blue sea.' " Although the Spanish devil had been defeated, the sea remained a difficult, often dangerous foe.[42]

## 6 Making War upon the Chart

· · · · · · · · · · · · · · · · · · · · · · · · · · · · · · · · · · · · · · · · · · · · · · · · · · ·

Strategy is the art of making war upon the map.

—Antoine-Henri, Baron de Jomini, *The Art of War*

The District of Columbia's Sons of the American Revolution met in Washington on December 28, 1898. It had been four months since the United States won its war with Spain. A contentious debate over ratification of the peace treaty, annexation of the Philippines, and indeed, the whole question of American empire was set to begin in the Senate. About two hundred Sons and their guests attended the flag-draped dinner. Toasts were made and speeches given. The evening's most anticipated speaker then stepped to the rostrum with a toast, "In ye time of peace prepare for war." He was Commander Royal B. Bradford, a naval officer and chief of the Bureau of Equipment whose administrative duties included oversight of the Hydrographic Office. Bradford had just returned from Paris, where he had served as an advisor to the Paris Peace Commission during the negotiations that ended the war with Spain. He had played an important role in the peace commission's deliberations and, in particular, the commissioners' intent to claim the entire Philippine archipelago for the United States. Now, returned from Paris and speaking before the gathering, Bradford enjoined his audience to his cause. "I am an expansionist," he declared. He was in good company. "We have already the most important station in the Pacific," Bradford declared, referring to the Philippines. "Let us keep it. As long as we don't own it, it will be a menace to us." He cited so many fine harbors he thought were ideal for coaling stations and followed this with numbers about the hopelessly short range of the battle fleet without stations to top its coal bunkers. "It is impossible for this nation to become an important naval strength without coaling stations all over the Pacific," he urged, echoing Mahan. When he had finished, the crowd roundly applauded and joined together in singing, "Columbia, the Gem of the Ocean." "The Army and Navy forever, Three cheers for the red, white, and blue."[1]

Two months earlier, Bradford had arrived in Paris to press the navy's strategic interests on the American commissioners, an argument he made primarily in cartographic and hydrographic terms. To defend a far-flung territorial empire, which now had begun to emerge from the aftermath of the war with Spain, the navy needed coaling stations and bases from which to operate. Access to coal for a steam-powered fleet—particularly in time of war—elevated logistical considerations to the fore. The questions remained, where would these stations best be located and what would that mean for the annexation of part or all of the Philippines? Bradford met the commissioners, "fairly filling my room up with the multitude of charts he had brought on from the Navy Department for use in his examination," observed Whitelaw Reid, a commissioner and Republican expansionist editor of the New York *Tribune*. Bradford's charts would be central to the way he articulated arguments in favor of American empire.[2]

He began with a primer in Pacific Ocean geography, constructing in the minds of the commissioners the visual space from which to imagine the expanse and boundaries of an American empire in the making. "Where are the Philippines?" asked William R. Day, former secretary of state and president of the commission. "Have you a map showing the American and Asiatic shores, both?" Thumbing through his charts, Bradford replied: "yes, here it is. There are the Ladrones [Marianas]; here are the Carolines; there are the Marshalls; here are the Hawaiian Islands; and there are the Philippines," pointing to the various island groups in the North Pacific on the line from California to China. "Here are the Pelews [Palaus]," Bradford continued, "about 600 miles from the Philippines. I am firmly convinced," he told the commissioners, "that the Pelews, Carolines and Ladrones should all be acquired." And so Bradford, speaking as a naval officer and one particularly concerned with the logistics of coal, expanded the scope of the commission's discussion. Annexation of the Philippines, he testified, required stations that spanned the Pacific, particularly in light of emerging German interest in staking claims to the vestiges of Spain's empire in the Caribbean and the Pacific. It was an argument that seemed plainly evident to anyone who understood the logistical limits of the reciprocating steam engine and then examined Bradford's charts.[3]

The depth of Bradford's hydrographic knowledge and its centrality to the political and military discourse of empire came into focus as the commission turned to the Philippines. Bradford contended that Palawan, a string bean–shaped island running from Mindoro to Borneo, with a commanding position on the South China Sea, was perhaps the best strategic position in the Philippines. It was his choice for a naval station in the islands. "There are five bays with good anchorages at any time or with any wind," he stated, which, he thought, made it "sufficiently valuable to excite the cupidity of any nation." But it was Bradford's own cupidity that was most excited. Referring to Palawan's Malampaya Sound, he launched into the hydrography of the matter:

> It is 19 miles deep, with a width of from 2 to 4 miles. The entrance is six-tenths of a mile wide and between bold and high headlands. It has been aptly named 'blockade strait.' The sound is divided into two parts of about equal depth. The channel to the inner section passes between islands, commanding the approaches and affording the most perfect means of defense. Within is a broad sheet of water, from six to ten fathoms deep, affording excellent anchorage and good holding ground. The entire sound is surrounded by high lands, is well wooded, and affords an abundance of good water.

Bradford thus couched his arguments in a new strategic language of hydrography—that is, he articulated his larger vision of American empire to the commission using environmental arguments about the advantages and disadvantages of particular bodies of water, which, like "blockade strait," imposed military meanings and uses on the natural world.[4]

At the turn of the twentieth century, American naval science militarized a marine environment framed by charts and a discourse of hydrography in which the Americans drew on environmental arguments to articulate the military advantages of a natural world now seen through a new, strategic lens. The imperatives that had driven American commercial empire were reduced though not eliminated, eclipsed now by a more muscular ocean empire, cast in the shadow of Mahan's vision of sea power, and lent urgency by the hydrographic difficulties the Americans had faced during the war. Naval science itself was militarized in ways that further removed it from the mainstream of civilian marine science, with which it had a long, but conflicted history. Absent from this new hydrographic discourse were surveys of indigenous cultures such as those that had animated the voyage of the Ex. Ex. Gone was sustained theoretical or oceanographic study of the ocean such as that which had framed Maury's *Physical Geography of the Sea*. Gone, too, was the sort of collaborative work with civilian science that defined Sigsbee and Agassiz's mutual, if diverging interests in the deep sea sounder aboard the Coast Survey steamer *Blake*.

The navy now became fully committed to studying the strategic qualities of the ocean and the bays, harbors, and coasts that now promised to be flash points in future wars and contests in defense of the American empire.

In the coastal waters that fringed the marine environment of this new empire lay what environmental historian Chris Pearson has called "militarized landscapes," which he defines as "simultaneously material and cultural sites that have been partially or fully mobilized to achieve military aims." In a similar way, the Navy's new hydrographic charts conjured a militarized seascape framed in postwar surveys of Cuba and the Philippines and the war games, or so-called chart maneuvers, in which officer students at Newport's Naval War College navigated diminutive battle fleets across a scaled, two-dimensioned sea and then took their newfound tactical and strategic understandings from Long Island Sound to Cuba and the Philippines, viewing these waters now with new eyes attuned to the strategic nature of the natural world. Charts and this language of hydrography appeared in public spectacle and ascended to the highest levels of the navy's strategic decision making in the period between 1898 and 1903. In them, we can trace not only the emergence of a new empire but also the ways that empire came to be defined, and even limited, by the natural world and the Americans' attempts to know and control it.[5]

In the fall of 1901, a court of inquiry examining Rear Admiral Winfield Scott Schley's conduct at the Battle of Santiago de Cuba, July 3, 1898, brought the chart and its inadequacies to the public's attention and pointed to the ways a broader strategic cartography came to underscore American sea power in the period immediately following the war with Spain. At the outset of the naval battle, which proved to be a decisive moment in the war, Schley had ordered a turn to starboard, which brought his flagship, the cruiser *Brooklyn*, into a near collision with the battleship *Texas*. This turn impeded both ships' pursuit of the fleeing Spanish fleet and momentarily called into question the outcome of this otherwise lopsided battle. The court formed one episode in the divisive Sampson-Schley Controversy over whether Schley or Rear Admiral William T. Sampson, in overall command of the North Atlantic Fleet, deserved credit for an American victory that, for all intents and purposes, ended the war and propelled the United States into the first rank of imperial powers. The court of inquiry was a public spectacle. Observers packed the courtroom at the Washington Navy Yard. The press offered daily excerpts of testimony as witness after witness recounted the dramatic opening salvoes of battle. Thomas Edison created two early motion pictures, one showing Schley astride the *Brooklyn*'s bridge, confidently directing the fleet, and Sampson, absent from the battle, sharing tea with some obsequious women. In defense of his client, Schley's counsel, Isidor Rayner, drew the court's attention to the *Navigator's Chart*, which

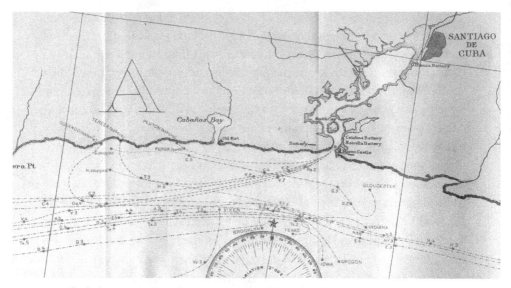

Detail of *The Navigator's Chart*, Battle of Santiago de Cuba, 1898, from the appendix to the *Annual Report of the Secretary of the Navy for the Year 1898*. The chart was a central piece of evidence in the court of inquiry examining the conduct of Rear Admiral Winfield Scott Schley. Courtesy of the Navy Department Library, Naval History and Heritage Command.

identified the positions of the American fleet in a blow-by-blow cartography of the American victory.

Rayner referred to the chart in his cross-examination of *Texas*'s navigator, Commander Lewis C. Heilner, who had a hand in constructing the chart, hoping to discredit Heilner's claim that Schley's order imperiled the flagship and his own. For Rayner, the chart was supposed to be authoritative. "This is something signed by you and the other navigators and ordered by Admiral Sampson, and returned by the Secretary of the Navy to the Senate, and the work was done less than three months after the battle," Rayner pressed. Exasperated, Heilner demurred. "But we are going over the whole business on a chart that is absolutely worthless—." Rayner had his witness where he wanted. "That is what I want to get at," the attorney replied, "that that chart is worthless—." And so there was agreement, at least for a moment, among Schley's defenders and his detractors that a chart of battle, officially appended to the secretary of the navy's annual report of 1898, was significantly flawed.[6]

Indeed, anyone closely following the proceedings at the Washington Navy Yard might be excused for concluding that naval officers were poor chartmakers. As spectators packed the room and newspapers offered daily excerpts of the proceedings, chart after chart was dismissed for its inaccuracies. "The public is taking a deep interest in the matter," reported the *Register* of Mobile, Alabama, specifically citing "incorrect charts, unreliable log records, failure to transmit orders, alteration of official telegrams and reports." At one point, Captain James Parker, another of Schley's counsels, questioned the use of a chart from the Hydrographic Office, issued to the American fleet in the recent war, but whose "positions . . . are grossly inaccurate. The coast line is put on that chart as 6 miles farther south than it ought to be and 4 miles farther west.". Indeed, given the Navy's experience in 1898, the court and its spectators had reason to question whether the charts were accurate reflections of the battle and the surrounding waters, or even whether they could be depended on as evidence in these proceedings.

As Rayner returned to the *Navigator's Chart* and to Commander Heilner's testimony in his closing argument, he attempted to reconcile distances put down on it with the navigator's testimony of how close *Texas* and *Brooklyn* had come to collision in battle. "It appears to me," Rayner concluded, "that every navigator was trying to put his ship in a different position from where his ship really was." He was careful to draw out contradictions in Heilner's story. Heilner had testified to 150 yards as the distance between the ships. But the chart—the one that Heilner himself had helped to construct—showed the distance between the two ships to be a safe 2,400 feet. "Here is a gentleman who, if it was true, would have liked very much to have placed his ship either right underneath the *Brooklyn* or right on top of her," Rayner argued, "on this chart he puts his ship 2,400 feet away at the time of the supposed danger of this collision." The chart may have been worthless, but it had exposed the contradictions in Heilner's testimony and, Rayner believed, proved the righteousness of his client as well. However faulty, it was all that stood between conflicting testimonies and the truth of what happened on the day of the navy's great Mahanian victory. As Rayner reemphasized, the chart "was the only thing that was given to us."[7]

Yet despite all the abuse heaped on the much-maligned *Navigator's Chart*, Rayner could not dismiss its power. "It is a remarkable document," he told the court. "Here are a half dozen navigators who meet together for the purpose of giving to the country a chart of the battle of Santiago, and after three months . . . they compose a chart which might as well be a chart of the battle of Salamis or the battle of Thermopylae or of the field of Waterloo." To Rayner, the United States seemed a new Athens or Great Britain—not coincidentally, the preeminent sea powers of their day—and Schley, a modern Leonidas or Wellington. In Rayner's mind, the chart itself seemed to evoke the gravity of the events it represented, a testament to American naval power and given to the nation, in Rayner's words, as a commemorative of one of the most decisive battles in the history of the Western world. Arguably no other sea battle—with perhaps the exception of Manila Bay, which began the war—was as important in the evolution of American empire. Despite its well-publicized flaws, the *Navigator's Chart* and the momentous events that seemed to play out across its dimensions had entered the public's imagination through Schley's court of inquiry. While it did not ultimately save Schley from censure, the chart both raised questions about cartographic accuracy and, at the same time, bore the symbolic weight of representing the decisive moment of American sea power at its inception. As Raynor suggested, it seemed to reify the emergent place of the United States on the world stage.[8]

As Schley's court of inquiry captured the imagination of an American public breathless with victory, the navy turned to more prosaic, but still significant matters, namely the logistics of defending the new American empire. Establishing coaling stations to succor the fleet and bases of operations in future wars were the foremost strategic questions the navy faced in the postwar era. "It is almost as difficult in the present day to exaggerate the importance of coal as it is that of air or water," remarked Captain French E. Chadwick, a veteran of the Cuba blockade and incoming president of the Naval War College. A fleet of modern, steam-powered warships, operating far from American shores, required bases and coaling stations, which the United States did not have. "The subject was forced upon the attention of the [Navy] Department by the Spanish war," Bradford had written, referring to the logistical as well as the larger strategic frustrations the navy had experienced in 1898. As chief of the Bureau of Equipment charged with coaling the fleet, Bradford was forced to pull ships from an already-thin blockade to coal at Key West or by collier at sea. It was only after the Marines seized Guantánamo Bay that the logistical challenges of coaling American forces began to abate. After the war, Secretary of the Navy John D. Long concluded that the service "found itself greatly hampered by the lack of coaling stations both at home and abroad." As Bradford's role in the deliberations of the Paris Peace Commission illustrates, just where to establish those stations and bases became a matter of hydrography as much as politics, diplomacy, and geography.[9]

The marine environment mattered to broader strategic discussions of American empire after the war, and the navy from Bradford down to the hydrographer of the navy recognized both the flaws of American hydrography and its implications for American sea power in the new century. The war with Spain exposed hydrography as a strategic weakness that had hindered American command of the sea. At Paris, in the course of the Philippines debate, Senator George Gray, the only Democrat and antiexpansionist on the commission, asked Bradford about the approaches to Palawan. Bradford admitted that his choice for a naval base was "more or less fringed with shoals, rocks and islets, making navigation dangerous with the present charts in places. The Philippine Islands are not well surveyed," he confessed, "and it is unknown dangers that are most feared." It was much the same throughout the new American empire. "From personal experience," reported Commander Chapman C. Todd, the navy's new hydrographer and a veteran of the Cuba blockade, "I am well aware that the same condition exists as to the published charts of the waters around the island of Cuba. What has been said relative to the Philippines applies equally to . . . Cuba." Hydrography, Todd now believed, had become both "a matter of commercial supremacy and national security." The navy's charts were too inaccurate to secure command of the sea, and they were inadequate to inform the debate over where to establish naval bases and stations. It was primarily in the context of these new strategic needs that the navy began to survey the waters of the American empire after the Spanish-American War.[10]

In some ways, it might be said that the strategic importance of hydrographic surveying and chartmaking became to the Americans what it had long been for the Royal Navy, whose hydrographic work had underwritten the spread of British commercial and naval power throughout their empire in the eighteenth century and into the nineteenth. It was through a "marriage of science and navigation," historian Michael Reidy argues, that "Britain began to control the waves and many of the world's colonial possessions." Still, the charts produced by the U.S. Navy did not, even by the turn of the twentieth century, equal the Royal Navy's efforts in breadth and scope, but neither did its new territorial empire rival Britannia's. This American hydrography need not match Britain in order to be expansive in its claims and potent in the ways it reframed visions of America's commercial and strategic role on the oceans. At the same time, hydrography never occupied a sustained prominence within a navy still circumspect about science within its ranks. The U.S. Navy never achieved the wedding of naval and civilian science that seemed to coexist more easily in the Royal Navy. From Fiji to Mexico, Hawaii, and Cuba, America's political and social embrace of empire was fraught in ways not mirrored in Great Britain. Even the new American empire that emerged after the war with Spain represented a short burst followed by a long rehashing and reappraisal of America's global commitments from anti-imperialists and isolationists. Empire was a different animal in the United States. The American navy, for all its unabashed emulation, was not the Royal Navy. This American cartography of empire as it evolved through the nineteenth century and into the first decade of the twentieth was potent and expansive, sometimes meteoric, but also stifled, and limited, as I have argued, by the inherent limits of cartography in a vast, dynamic marine environment and, of course, by many other political, social, and cultural factors ranging from anti-imperialism to the relative embrace of science within a service that still remained largely unscientific in its culture, values, and worldview.[11]

Between 1898 and 1903, as naval officers considered establishing the bases and coaling stations that Mahan had been writing about for more than a decade, they looked to the sea as an environment that needed to be understood within an emergent strategic framework. Mahan seemed to grasp the hydrography of the matter in the fall of 1898 when, as a member of the Naval War Board that advised Secretary Long, he called for surveys to inform the base debate. "Before any harbors that may be selected as naval stations are permanently acquired," the board counseled in a report to Long, "each should be visited, carefully examined and reported upon fully by competent naval officers sent for this purpose in one of our cruisers." If Mahan did not quite grasp the transoceanic scope of the work in proposing to dispatch a lone cruiser here and there, the General Board of the Navy did. Established by Long in 1900 as a permanent advisory council to the secretary, the General Board spent the better part of its first three years considering the hydrographic depths of naval strategy. George Dewey, hero of Manila Bay and now Admiral of the Navy, served as its president. In the 1870s, while commanding the steam frigate *Narragansett*, Dewey had directed a three-year hydrographic survey of the Pacific coast of Mexico. In 1901, Dewey wrote to Long that "as complete a knowledge as possible should be possessed by the Board, concerning the hydrography of such points as may be utilized for naval bases or are of strategic value to our naval forces in the event of hostilities with a foreign naval power." As Dewey put it in one General Board report, "without knowing the bottom thoroughly we lose the benefit of it." At the same time, officers at the Naval War College, the navy's center of intellectual and strategic thinking, were also grasping hydrography's newfound significance. "There is an enormous amount of hydrographic surveying to be done in our new possessions, owing to their multitudinous insular character and the meager and imperfect hydrographic work already done in them," declared Captain Charles H. Stockton, president of the Naval War College, when he appeared before the United States Senate in 1899. There was thus a broad understanding from the highest levels of naval strategy and decision making in the General Board through Bradford at the Bureau of Equipment and the hydrographer of the navy in Washington, that further surveys and chartmaking would be central to the expansion and defense of American empire.[12]

Bradford cast hydrography in the language of sea power, part and parcel of this new age of naval warfare. Underscoring hydrography's military significance, he reported to the secretary of the navy in 1903 that "there is no more important and necessary work with a view of being prepared for war. In the opinion of the Bureau it is quite as necessary for this Government to be able to supply to its ships of war all the charts necessary for purposes of navigation anywhere in the world, as it is to supply them with armor, ordnance, coal, and other articles of equipment." To Bradford, charts ranked with steel and shot in the arsenal of naval war. Indeed, by 1900, both Bradford and the General Board had concluded that the Navy embark on a system of confidential charts, retaining for the Navy's own exclusive use hydrographic knowledge of strategically important waters rather than the traditional practice of sharing navigational knowledge with all. That he equated their importance with coal—his great passion as bureau chief—says a good deal about the new role hydrography played in naval affairs. Congress obliged, significantly increasing the annual appropriation for ocean and lake surveys in the Hydrographic Office's budget from $14,000 in 1898 to $100,000 in 1899. It remained there through 1905 and at $75,000 for the next four years. With hydrographic considerations at the fore, Bradford at the Bureau of Equipment and Commander Joseph E. Craig at the Hydrographic Office planned a comprehensive survey of Cuba, which Bradford considered among his most important work at the bureau. Cuba, of course, had been the site of the navy's greatest hydrographic difficulties during the war. The island's position astride the Windward Passage—a stretch of water separating the island from Hispaniola and commanding the main sea lane to the isthmus—assured its strategic importance to a canal. Any major American naval base in the Caribbean would almost certainly be located in Cuba. The only question remained where.[13]

In January 1899, the yachts *Eagle* and *Yankton* began hydrographic surveys to answer that question. With the addition of the yacht *Vixen* late in 1899, the three survey vessels worked in Cuban waters for the next four years. The three had been veterans of the Cuba blockade, acquired by the navy from private owners in the first weeks of the war. They were steel-hulled, both sail and steam-powered, and they were small—less than two hundred feet long and about twelve feet in draft—making them ideal for surveying in coastal waters. On their coming work and countless other hydrographic surveys across the new territories largely rested the particular geography of American empire anchored by major American naval bases in the Caribbean and the Pacific whose locations were yet to be determined. If the war had opened the question of American territorial empire and framed its broad geographic contours, and if diplomacy such as the Paris treaty and the Platt Amendment of 1901 had set the scope and nature of annexation in the Philippines and intervention, for example, in the affairs of Cuba, it was hydrography and the strategic nature of the marine environment that underscored how and from where the Americans would defend the empire and project sea power across the ocean.

Though there were some differences, hydrographic surveying in the new century remained much the same as it had when the United States Exploring Expedition charted the Fijis in 1840. The differences were largely technological. Instead of sail and oar, the launches that set out from *Eagle*, *Yankton*, and *Vixen* were steam-powered, making the intricate maneuvering demanded by triangulation much faster and easier. In addition, the entire survey could be plotted more accurately since longitude, long the most complex and inaccurate of navigational measures, could now be determined more precisely by its relation to telegraph stations whose coordinates were known by signals sent from Cuba to the prime meridian at Greenwich. But the method of the trigonometric survey remained the same as it had in the 1840s.

The whole exercise remained subject to the vagaries of nature. The surveying schedule itself was interrupted by a four-month break during the rainy season when hurricanes and lesser storms battered the Cuban coast. The island's shallows were so vast and complex that the surveyors often had to work almost out of sight of land. Even close-in, the mangrove swamps that surrounded many harbors made the erection of signals difficult. Aboard USS *Yankton* in Nipe Bay, Commander George Dyer wrote his wife "fearing the unscrupulous natives . . . will have despoiled our signals," which along with "the strong winds to date," convinced him that "much work will be required to place things in status quo ante." The tropical sun continued to do its worst, glancing off the exposed coral heads in a blinding panorama of light that had been as familiar to Passed Midshipman William Reynolds in the Old Navy as it was in the New. Ensign Ernest J. King, who, during the Second World War, would rise to become Chief of Naval Operations, temporarily lost his eyesight while surveying aboard *Eagle*. "I am sure that none of the cadets would be able to use a sextant after a continuous week of this sort of sounding," Commander Carlos G. Calkins, *Vixen*'s commanding officer, reported to Bradford. "Already we have many complaints of damaged vision." Much of the work was "done in water rough enough to make Midshipmen seasick," remarked Lieutenant Commander M. L. Wood in *Eagle*. With higher rank perhaps came sturdier constitutions, but the fact remained that this work, like no other in the navy, immersed officers and men in the marine environment.[14]

Bradford intended first to locate a suitable site for naval use, and so he directed *Eagle* and *Yankton* to proceed to Guantánamo and Santiago bays, respectively. Geography had placed these harbors favorably, commanding Cuba's southeast coast and guarding the vital Windward Passage leading to the approaches of a yet unbuilt canal. Other harbors such as Havana and Nipe Bay on the north side of the island and Cienfuegos on the south were also considered, but, for the navy's needs, they were not close enough to this most strategic stretch of water. "Opinions are divided as to which is the most desirable port, Santiago or Guantánamo, for deposits of coal for naval purposes," observed Bradford. The choice would primarily be a hydrographic one. By the time both vessels returned north to spend the rainy season of 1899 in cooler waters, they had accurately charted both bays. At Santiago, whose prominence as a commercial port meant that it had been better surveyed, *Yankton*'s attention focused on charting the narrow, circuitous channel that was Santiago's dominant strategic feature. At Guantánamo, forty miles to the east, *Eagle* surveyed the bay's more than one hundred miles of coastline and twenty-five square miles of water. Its crew erected more than two hundred signals and took over twenty-five thousand soundings. The survey found a new deep water channel not on the old chart, which had been unknown to the local pilots. With little dredging, the channel would allow battleships to coal closer to shore, one tactical advantage in wartime that likely pleased Bradford. Otherwise, he summarized, "important shoals were more correctly located and developed and the hydrography was corrected in many places." By the summer of 1899, Guantánamo and Santiago were as well charted as any harbor of the continental United States.[15]

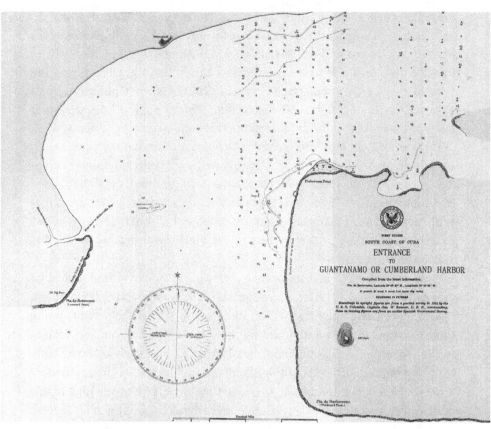

Detail of *Entrance to Guantánamo Bay or Cumberland Harbor,* from Spanish and American surveys, 1898. This was the extent of hydrographic knowledge for Guantánamo during the Spanish-American War. Note the lines of soundings and compare to later efforts (facing page). Courtesy of the Library of Congress, Geography and Map Division.

Between 1899 and 1903, the three yachts encircled Cuba with new charts that served both commercial and military purposes. Though strategic imperatives had transformed the navy's hydrographic work and its cartographic thought, the broader economic underpinnings of empire articulated by historian Walter Lafeber and others remained. Though primarily interested in the island's strategic characteristics, the navy and its Hydrographic Office still provided nautical charts to the nation's mariners and, increasingly, to an American business community with overseas investments. American companies that were now drawn to invest in postwar Cuba demanded better charts, and so the Hydrographic Office remained beholden to commercial interests. Lieutenant-Commander Frank Friday

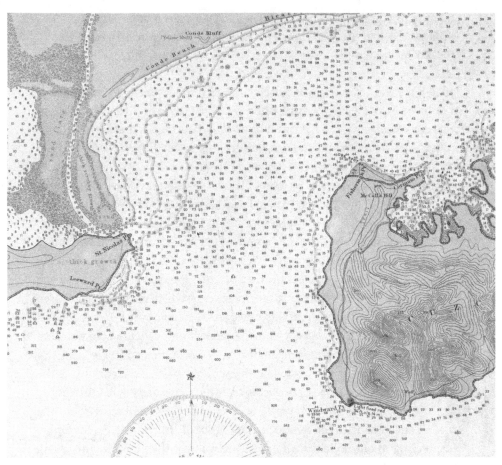

Detail of *Entrance to Guantánamo Bay*, 1900. This chart, produced by the navy after the Spanish-American War, reflected the survey of USS *Eagle*. Note the extension soundings in the entrance relative to earlier efforts (facing page). Courtesy of the Library of Congress, Geography and Map Division.

Fletcher, who commanded *Eagle* in 1900, informed General Leonard Wood, governor-general of the island, that "considerable capital is being invested along the N.E. coast by Americans and I have reason to believe that other capital is being held back owing to the lack of more definite information relative to the coast." Fletcher hoped to enlist Wood's influence to have the navy's hydrographic surveys continued on the north coast of the island. Fletcher told Wood that New York's Munson Steamship Company, which sought to develop the harbors of Nuevitas and Gibara, was "unwilling to invest their money in those wharves and improvements until more definite information is obtained as to depth of water, character of bottom, shore line and other facts embodied in a modern plan and hydrographic survey." Thus, while the sea had become a strategic domain, the navy by no means wholly ignored continuing commercial needs. The two were, in fact, often inter-related.[16]

Still, it was the extant commercial values implicit in the existing charts of Cuba and the Philippines that often rendered them of little value in strategic discussions or in navigating a blockade along coasts that were not previously significant to maritime commerce, but nevertheless would become strategically and tactically important in time of war. The hydrographic concerns of American business were not always commensurate with American naval strategy. While the navy considered operating from the well-charted harbors of Havana, Cuba, and Manila in the Philippines—long the centers of maritime commerce in Spain's empire—the Americans ultimately found neither to be geographically or hydrographically suited for an American naval base established to protect the Caribbean approaches to a canal in the case of Cuba or defense from an attacking enemy, which naval leaders feared at Manila. Military considerations, in contrast to commercial ones, generally informed the Cuba survey, and a succession of officers commanding these yachts viewed the marine environment through a strategic lens.

If the Cuba survey had made one thing unquestionably clear, it was the uselessness of so many existing charts, which the navy had distributed during the war. The battle reports, of course, had attested to this, but the surveys revealed the true measure and extent of this flawed cartography. Responding to one solicitor who hoped to interest the navy in purchasing land in Cuba for a coaling station, Bradford responded, "with the exception of a number of ports surveyed by this Department since the Spanish war, the charts, generally speaking, of Cuban waters are so inaccurate that it is impossible at present to judge of the strategic value of the harbor mentioned by you." In the course of his survey, Commander Austin M. Knight commanding the survey vessel *Yankton* found his H.O. Chart 1523 "seriously in error." It had been constructed, he discovered, from a survey by the Spanish navy "apparently made by running a few lines of soundings and then sketching in the coast and cays by eye." Such inattention to hydrographic practice was anathema to the Americans, who, at least in theory and cartographic ideal, were committed to precision and comprehensiveness in surveying method. Distances erred by as much as five and a half miles. Aboard *Eagle*, Fletcher found that Guantánamo's position diverged from the chart by more than one mile of latitude and one of longitude. Bradford went so far as to suggest that the inaccuracy of Spanish hydrography had perhaps itself been a strategic act—that "the former government . . . preferred that the coasts of Cuba should remain a danger . . . rather than that accurate surveys should facilitate approach in time of war." In all likelihood, accurate hydrography of Cuba's coasts and harbors was the least of Spain's military concerns, but such was the ease with which he now found it to be a strategic asset that he could entertain the idea.[17]

Still, the Spanish had done an extraordinarily poor job of surveying their empire, something that the Americans hoped to rectify not only in the Caribbean but also in the Philippines, where the search for a suitable naval base was constrained by a whirlwind of political and strategic complications. In his first annual report as hydrographer in 1900, Commander Chapman C. Todd referenced the "inaccurate charts of the Philippine group." Indeed, the cruiser *Charleston* and several other ships and gunboats on the Asiatic Station had been lost or laid up due to groundings. Todd intended to construct a new set, which, he promised a fellow officer, would be "carefully thought out from the strategic as well as the navigational standpoint." At any rate, with regard to a naval base, Todd wrote, "until there has been a thorough examination of the more important points . . , I do not think anyone ought to express a final point on this matter." But the American position in the Pacific and the Far East was far from clear, making surveys comparable in scope and comprehensiveness to those off Cuba out of the question. Early in 1899, a Philippine independence movement took up arms against the new American colonizer. True to Bradford's word at Paris, Germany had purchased territory from Spain in the Mariana Islands, adding to their outposts at Samoa and at Kiaochow on the coast of China. Then, in 1900, an anticolonial Chinese uprising besieged the foreign legations at Beijing, the "Boxer Rebellion." The navy's Asiatic Squadron, in other words, had its hands full with pressing matters. Hydrographic surveys were necessarily a secondary concern. Still, the navy took up the question of a naval base in the Philippines, aware that Asia was an ever-growing sphere of economic and strategic interest and that the Philippines, as American territory, would require defense rooted in a better understanding of the natural world.[18]

In 1900, the Asiatic Squadron was perhaps the most demanding command in the navy, but it nevertheless fell to Commodore George C. Remey to lead a reconnaissance of Philippine harbors amid his other duties as a wartime commander. At the behest of Congress and Secretary Long, the Remey Board met for the first time aboard the commodore's flagship, the cruiser *Brooklyn*, on December 10, 1900. Its work was limited by the

exigencies of war, the urgency of time, and the vastness of the archipelago, which consists of some seven thousand islands. Nevertheless, this
board of four naval officers and a civil engineer completed a rough hydrographic survey in thirty days. In its report to Long, the board dismissed
Malampaya Sound on the island of Palawan, which Bradford had pushed
when he appeared before the Paris Peace Commission. "From the indica-
tions on the charts and from other information," it concluded, "though
there are good anchorages . . . it is not the most suitable location for the
principal naval station in the islands." The board, continuing on, found
channels "practicable . . . for vessels of the deepest draft" at Iloilo in the
central Philippines—another site favored by many naval officers.[19] But it
unanimously selected the port of Olongapo in Subic Bay on the island of
Luzon, thirty miles from Manila. On January 8, 1901, Remey cabled Long,
citing Subic's "good channel, ample anchorage" and its "inner basin well
sheltered from storm waves," which, he added, "requires some dredging."
The official report followed. In its main points, the Remey Board echoed
the arguments that were at that same time being made in favor of Guantánamo. So the navy, which would ultimately choose both Guantánamo and
Subic to establish American sea power in the Caribbean and the Pacific,
cited common hydrographic features among its most important considerations and its determining factors.[20]

Hydrographic reports from the Philippines arrived intermittently, spurred
by the ongoing strategic work of the Naval War College and the General
Board's intent to glean as many opinions as possible before making its final
decision on the location of a base. At the War College, President Stockton
hoped to have some homecoming officers assigned to the staff. Lieutenant
John M. Ellicott had been a student at the college in 1896 and returned to
Newport once again from the Philippines to lecture on the strategic features
of the Pacific. Lecturing before the class of 1900, Ellicott dismissed Manila.
"From its topographic and hydographic environment," he declared, it was
"absolutely indefensible." Among other weaknesses, its entrance was too
wide to be protected and too deep to be mined. Contrary to the Remey
Board, whose findings appeared only a few months later, Ellicott favored
Iloilo. Its bluffs, he told the college's students, "drop back into a semi-circular
bight where thirty battleships could lie absolutely concealed from the outside." He continued, conjuring a strategic environment sprawling with dry
docks, repair shops, a coal station, and a supply depot—a naval base, in his
words, "preeminently the strongest to be found in the Philippines." But the
next year, Rear Admiral Frederick Rodgers, who succeeded Remey in com-

mand of the Asiatic Squadron, threw his influence once again behind Subic, whose "natural features" and harbor he too thought "magnificent." It was "without a question the most desirable in the Philippine Islands for a naval station." Subic, Iloilo, Palawan, Manila—Philippine harbors were tossed around the board rooms, classrooms, and ward rooms of the navy in these years with little consensus. This was certainly the result of the hastiness of wartime surveys in which Filipinos sometimes took shots at the surveyors themselves. But disagreement also stemmed from the fact that no harbor was perfectly suited in an environmental sense. Each presented a mix of strategic advantages and disadvantages.[21]

By the spring of 1900, the navy's survey work in Cuba and the Philippines placed the Hydrographic Office squarely in a political dispute with the Coast and Geodetic Survey over hydrographic jurisdictions and budgetary appropriations that hinted at deeper scientific disputes as well as the ambiguous place of these coastal waters within the political and legal frameworks of American territoriality. In some ways, the hydrographic domains of these two cartographic institutions neatly divided where the three-mile limit of American territorial waters met the rolling swells of the open ocean. In reality, however, the line was as fluid as the sea itself. In Maury's time, for example, the navy and the Coast Survey both claimed the Gulf Stream as their hydrographic purview. After the Civil War, both the navy and the survey set out to chart the new Alaska Territory. Now, however, the ambiguity over the extent to which Philippine or Cuban waters would become American territory added a new dimension to this long-simmering scientific and political rivalry, particularly in renewed attempts by the navy to abolish the Coast Survey and merge the nation's hydrographic institutions under naval administration.[22]

The crux of the navy's rationale for usurping the Coast Survey's work in waters that, in many cases, were in the process of some as-yet undetermined transformation into American territorial waters was that the coasts of Cuba and the Philippines was strategically important in ways that the civilian institution would never understand. The Philippines were then an active war zone. Naval officers seemed best suited to see the marine environment and then chart it through this lens. "We, who have to go to sea in the vessels, know the immeasurable value of accurate charts," Commander Todd, the hydrographer of the navy, wrote to one of his subordinates. His superior (now promoted), Rear Admiral Bradford, echoed his chief hydrographer. "The experience gained will diffuse through the naval service a knowledge of the waters surveyed, which will act as a measure of security in the navigation of the vessels of the Navy." Knowledge of the marine environment was fundamental to the naval officer's profession, according to Todd and Bradford. A grounding might imperil his professional career and, perhaps, alter the course of battle. As Congressman George E. Foss, chairman of the Naval Affairs Committee, put it on the House floor during an appropriations debate, "when you expect [naval officers] to know every rock, and every reef, and every shoal . . . you ought to give them the right to make the surveys for the uncharted seas."[23]

To the Coast Survey and its supporters in Congress, the navy's claims smacked of militarism, of an expanded role for the navy outside its political, legal, and scientific domains by naval officers who, though they might fancy themselves scientists, were nevertheless by the late nineteenth century outsiders in a more clearly defined scientific profession. "Why, gentlemen, we must not expect these naval officers to devote their hours of retirement from active duty to work in which they have no interest and work for which they have no special fitness, but only the fitness that is possessed by well-educated, cultivated men," declared Congressman William H. Moody during the House debate on cutting appropriations for the Hydrographic Office's ocean and lake surveys. "It is the same old story of the expert professional against the amateur," echoed survey superintend Henry S. Pritchett, an astronomer and future president of Massachusetts Institute of Technology, in testimony before the Naval Affairs Committee. During the antebellum era, such arguments would not have been unfamiliar to Maury, whose rivals in science, Joseph Henry and Alexander Dallas Bache, had leveled similar critiques at the superintendent of the Naval Observatory. Old rivalries in federal science emerged anew, recast within the larger stakes of American empire. Together, these bureaucratic turf battles marked the growing divide between naval science, which had largely abandoned larger theoretical questions of marine science for the strategic work of imperial defense, and the Coast Survey, which joined the United States Fish Commission, the Marine Biological Laboratory at Woods Hole, Massachusetts, and the Marine Biological Association (later Scripps Institute of Oceanography) in pursuing broader questions of marine science and oceanography.[24]

The navy's encroachment on Coast Survey duties incited the ire of some in Congress, where a prolonged appropriations battle briefly brought this scientific feud to the public's attention and pointed—in its debate on the hydrographic jurisdictions of these two organizations—to just how ambiguous American claims to these coastal waters were at the beginning of the new century. In April 1900, as Congress debated major cuts in the Hydrographic Office budget for "Ocean and Lake Surveys," assuming that such work in American territorial waters could be done by the Coast Survey, Commander Todd, the hydrographer, was temporarily relieved of command for circulating a letter to branch hydrographic offices across the nation encouraging the merchant communities in those cities to lobby their representatives on behalf of the navy in the matter. Meanwhile, Congress attempted to parse whether the coastal waters of Cuba and the Philippines were American territory. "Gentlemen I am sure will agree," Representative Joe Cannon explained, "that Cuba is not a part of the United States. Some perhaps think so. I do not say they are. Others are under the impression that the Philippine Islands are not part of the United States. Others think that they are, but ought not to be; others say they are and ought to be. There is a manifest difference of opinion on that question."[25] As the American empire emerged in the rhetoric of political debate and in the American mind, these questions of hydrographic surveying in American territorial waters laid bare the messy process of incorporating these islands and their coastal waters into the sovereign domain of the United States.

As the debate concluded, Foss of the Naval Affairs Committee rose to give what would prove a compelling argument in the navy's favor. "I want a military government maintained there for the present," Foss declared, referring to the Philippines. "I am surprised at gentlemen on the other side of the House who are against the retention of the Philippines and fight 'expansion' so vigorously, who do not believe that the civil government can go over there, and yet are voting for a proposition to extend the Coast and Geodetic Survey across the seas into those far-away regions." He was referring to anti-imperialist Democrats who nevertheless supported the Coast Survey in this very narrow political battle. But, according to Foss, this might be construed as much more. Assenting to the Coast Survey's hydrographic jurisdiction in Philippine coastal waters, according to Foss, would acknowledge that the Philippines were inseparably a part of the United States. As Foss acknowledged, the implications could well be profound:

That means, Mr. Speaker, if you send your Coast and Geodetic fleet over to make these surveys, it will be followed by other departments of the Government in order. It will be but a short time before the Geological Survey will go; the Land Office will go, and you will find every branch of the civil government as it is organized here gradually extending itself into the Philippine Islands. Now, I ask how gentlemen on the other side of the House can reconcile themselves to that condition of affairs? I ask how they, holding the views they do, can vote for a proposition like this? I would like to have them explain that question.

Permitting the Coast Survey to chart these waters committed the nation, as far as Foss was concerned, to a definition of these islands as a civil extension of the United States. Civil government led inexorably to a permanent American empire that at least some members of Congress were unwilling to sanction. Annexation had, in fact, not settled the imperial question; rather, it had raised difficult new ones.[26]

The political stalemate on this issue dissipated almost immediately with a tacit recognition that enough hydrographic work remained in the new empire to keep both the navy and the Coast Survey busy for years to come, but also that the meaning of these waters—strategic, political, territorial—was still very much contested. The very notion of American territoriality in the waters of the new American empire played out in an otherwise mundane congressional appropriations debate over hydrography, which nevertheless held extraordinary meaning for the growth of American national power in waters from the Caribbean to the Western Pacific.

As the navy continued surveying its new empire, the scope of environmental knowledge it collected and then debated between 1899 and 1903 was expansive, indicative of the centrality of the natural world to American imperial ambitions. As Mahan and Bradford knew, defense of the Philippines required not only a base in the islands, but a string of stations across the Pacific to support it. These, Ellicott had told the War College in Mahanian fashion, "must be considered like strongholds along a military highway," after which he delved into the hydrographic and strategic qualities of each. Here was Maury's common highway militarized in the service of fleet logistics and defense of the empire. The charts of Guantánamo and Santiago were precise to the thousandth sounding; in 1903 alone, *Yankton* made a remarkable 180,000 such measurements. In many cases, the charts were vastly superior to the prewar charts, many based on Spanish surveys, available to American naval commanders in 1898. Here was a hydrography of American empire that not only replaced incomplete charts with more accurate ones but recast the whole strategic environment according to American hydrographic standards of practice and in the service of American interests. In short, these charts helped transform what had been a crumbling Spanish empire into what the Americans hoped would be a new, more muscular one.

Throughout the empire, new surveys, new charts, and the natural world proved critical to the ways the navy imagined, thought, and talked about the new empire. If the Cuba charts were impressive in their scope, other surveys—of the Philippines, for example—were less thorough, and many more, from the coasts of China to Puerto Rico, lay somewhere in between as the navy made due with shortages in personnel and vessels to carry out its vastly expanded duties. After 1900, the navy moved from Cuba to consider the hydrography of Puerto Rico, Santo Domingo, Haiti, Mexico, and both coasts of the isthmus. In the fall of 1899, the collier *Nero* followed the deep sea sounding work of USS *Tuscarora* in charting a course across the Pacific for a submarine telegraph cable that would finally link communications between San Francisco, Hawaii, Midway Island, Guam, and Manila, and in the process, recorded the deepest depth sounding ever made up to that time in the Mariana Trench, southeast of Guam. In November 1900, the auxiliary cruiser *Yosemite* completed its survey of the harbor of San Luis d'Apra, Guam, before a powerful hurricane hit the island, driving the ship from the harbor out to sea, where it sank with a loss of five men. "It appears that a breakwater is not considered advisable," a survey board concluded in 1901, "on account of its great cost and the uncertainty of its resistance against storms." In the same year, a pair of gunboats under the command of Lieutenant Albert P. Niblack, who himself had been a lecturer at the War College, completed a survey of Subic Bay. By 1903, despite ongoing war and, as always, the challenges of survey in a dynamic marine environment, the navy had amassed a considerable stock of hydrographic knowledge about its new empire. Though much of it still remained poorly charted, officers from the naval War College and the General Board drew on the charts that did exist and the opinions of the surveyors themselves as they debated naval bases and planned for future wars.[27]

At the Naval War College in Newport, Rhode Island, junior officers' three-month course of study enlisted new cartographic forms to plan for and execute naval war and, more broadly, to press officer-students to view the natural world through a strategic lens. In 1893, Captain Henry C. Taylor became president of the college, beginning a tenure that naval historians consider one of the young institution's most critical moments. Established at Newport in 1884 by Rear Admiral Stephen B. Luce, the college trained officers in naval history, tactics, strategy, and international relations, attempting to apply the scientific principles of the day to the study of naval warfare. It was as lecturer and president of the college that Alfred Thayer Mahan had revised his lectures into *The Influence of Sea Power upon History*.

Mahan, his book, and his thinking were thus deeply ingrained in the college's culture. Bradford himself had been a graduate of the inaugural class of 1885, and many of the bright young officers in command of the Hydrographic Office or making surveys in the Caribbean and the Western Pacific had since studied there. During Taylor's presidency, however, the classroom moved to the waters of Narragansett Bay and Long Island Sound to study the strategic elements of the marine environment firsthand. Typically, Taylor and his staff assigned small groups of student-officers to spend several days, with chart in hand, studying the hydrography and topography of localities that figured prominently in each year's problem—a broader strategic question around which each summer's course was organized.[28]

These forays produced what Taylor called war charts, which forced officers to think about the environment's strategic relevance and became the basis for the college's system of war planning. Despite the navy's increasingly offensive orientation, it nevertheless remained concerned, both before and after 1898, with coast defense. For the War College, the harbors and coasts of the United States were as strategic as the waters of Cuba or the Philippines. During the summer of 1896, for example, in the course of studying a problem whose parameters were the defense of New York City and Long Island Sound, Lieutenant Commander J. R. Selfridge and Lieutenant Ellicott chartered a small steamer for a reconnaissance of Fisher's Island Sound off the coast of eastern Connecticut. "During this trip," the officers reported, "the chart was frequently consulted and the various inlets, rivers, and harbors specially examined as places of refuge for torpedo boats and mosquito fleets, together with the necessary docking facilities, while the Sound itself was studied for defense." The waters off Fisher's Island, the two officers concluded, would make an ideal base for small vessels to harass any enemy intent on attacking New York City from Long Island Sound.[29]

This was precisely the kind of thinking that Taylor and others hoped to promote with these studies, which considered the American coast as a strategic environment and turned the chart itself from a navigational aid into a weapon of war. The war chart, President Charles H. Stockton instructed the class in 1899, "should be based or made upon a hydrographic chart of the area under discussion" and should include "all the features that will enter into the attack and defense of an anchorage, harbor, bay or water area." Taylor intended it to be both "an exercise of the mind in the study of war" and "of inestimable value to our fleet in a moment of crisis." Thus, Taylor told his students, "the commander-in-chief has only to hand the chart to the officer he selects as the commandant of the fleet base and direct him to

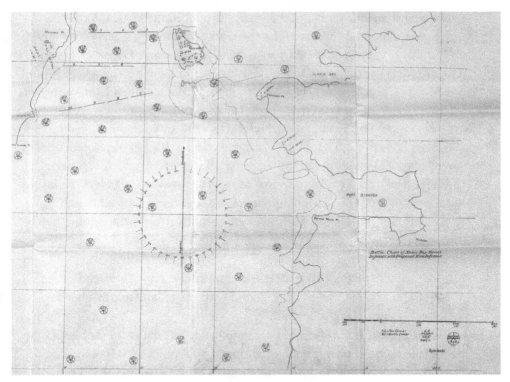

Detail of *Battle Chart of Subic Bay Naval Defenses* showing battery positions and minefields, indicative of the militarization of the marine environment and the ways a broader strategic cartography informed the navy's discourse, its chartmaking, and its conceptions of the marine environment in the new century. Courtesy of the National Archives and Records Administration.

carry out the details." The war chart, then, was a kind of military cartography, studded with symbols detailing fortifications, land batteries, anchorages, channels, and shallows where mines and torpedo boats might impede an attacking fleet, or as a place of rendezvous for the battle fleet to sortie against the enemy. As a visual representation, these charts were designed to be read and understood at a glance, when decision at sea could turn on a moment. Indeed, they were not unlike Maury's *Wind and Current Charts* in their visual power to convey complex information in a way that was immediately graspable. But where Maury's charts had led the mariner to favorable winds and currents, or fertile whaling grounds, Taylor's war charts intended to lead the naval commander to victory in battle.[30]

This militarized cartography also figured importantly in another exercise at the Naval War College—the war game. Lieutenant William McCarty

Little, a permanent member of the staff, introduced the game into the college's course in 1894, influenced by the works for the Royal Navy by John Clerk and Sir Philip Colomb, but also by the German army's *kriegspiel*, which had proved its worth in the Franco-Prussian War of 1870–71. The war game, like Mahan's histories and Luce's founding philosophy, intended to apply scientific principles to the study of naval war, so that, given certain parameters, ship duels, fleet battles, and naval strategy could be played out, tested, and retested, and principles of naval warfare then extrapolated. The game board, of course, was the chart itself. Assigned a particular scenario, the student-officers divided into two forces under two commanders to play out the game over a series of hours or days in separate rooms with nothing to govern their movements but the chart and directions from an umpire. This umpire kept abreast of the action and ultimately judged its outcome according to his own omnipotent chart. The hydrographic details mattered little here, but in matters of coast defense and attack, or, in the broad sweep of the strategic game, the environmental features of land and sea became inextricably part of this mock war.[31]

McCarty Little quite intentionally employed the chart as a representation of the tactical, operational, or strategic field of naval war, encouraging the officer-students to make imaginative leaps between the chart and future fields of battle, and, conversely, between the bridge of the officer's flagship among the smoke and fire of naval battle and the lessons learned earlier on the gaming chart. McCarty Little actually preferred to call the exercise a chart maneuver rather than a war game. The latter, he thought, "had much the same depreciating effect as the term Sham Fight has had with regard to field maneuvers." Chart maneuver, he told the students, "accentuates the fact that the strategist's real field of operations is the chart, just as the architect's real field is the drawing board." For McCarty Little, the war game or chart maneuver was meant to reflect reality, as close as it could be achieved on the chart. He then invoked the French military theorist Antoine-Henri, Baron de Jomini's well-known maxim from his military treatise, *The Art of War*, that strategy is "the art of making war upon the map." To McCarty Little, naval strategy was similarly making war upon the chart.[32]

In the naval officer's mind, the line between chart and sea blurred. Representation and reality merged. In fact, McCarty Little thought, the chart was perhaps a better representation than reality itself. "A little consideration will show that ordinary navigation is merely sailing on the chart," he argued. "A walk on deck gives no idea where the ship is, but a glance at the

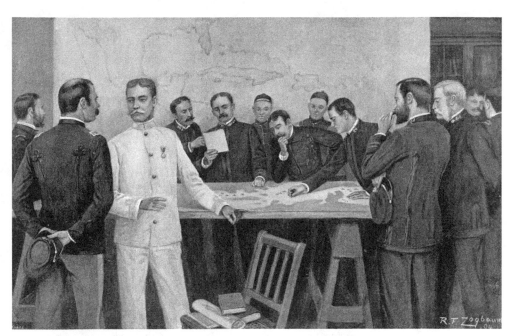

*A Problem in Naval Tactics—at the War College, Newport, Rhode Island,* by the artist R. T. Zogbaum, from *Harper's Weekly,* February 1895, showing officers at the Naval War College clustered around a war gaming chart with a map of Cuba, Puerto Rico, and the Caribbean in the background. Courtesy of the University of North Carolina Library System.

chart in the cabin does. In like manner it is on the chart that the admiral plans and conducts his cruise." McCarty Little had made the imaginary leap from representation to reality and then took the metaphor further. The chart was not simply a classroom stand-in for command on the bridge of a warship; rather, in battle, it actually afforded the commander a much broader vantage of strategic vision. "We must not overlook the fact that the game is a convention just as is the chart or printed page, or indeed language itself," he admitted, "and if we wish to use either, we must learn to think in it. The war game is a cinematographic diagram; and it is as important to us to be able to read it, as to read a chart or a book." War gaming was an imaginative act, one in which the game board or war chart mediated the relationship between the marine environment and the battlefield. In real war, then, the naval commander need hardly view the action so much as plot and then direct it on the chart just as he had done in the classroom. The war game was not altogether different from the battle itself. "Even on the tactical field with the enemy in sight, the picture on the retina is a distorted representation, which in the mind must be reduced to a proper diagram," he told his students. "Even the actual witnesses to a battle do not have a clear idea of what has taken place until it has been reduced to a diagram." In the officer's mind, the war game should become the war itself. With it, the chart became the sea—the commander's field of tactical and strategic decision.[33]

Naval officers took these new meanings cultivated in the course of the study at the War College and applied them to the waters of Cuba and the American empire more broadly. Commander Austin M. Knight, who had been a student at the Naval War College and would ultimately become its president from 1913 to 1917, is a case in point. In command of *Yankton* during the survey of 1902–3, Knight wrote to Bradford at the Bureau of Equipment and Dewey at the General Board regarding Buena Esperanza, an inland sea on Cuba's southern coast that had bewildered the navy in its assault on Manzanillo at the close of the war. After his survey, Knight found the waters no longer treacherous, but, in fact, ideal as an anchorage for the American battle fleet. "I know of no sheet of water in the world which is in any degree comparable to it," he observed, "nor am I able to think of any possible advantage in which it is lacking for the use of our Navy whether for drills in time of peace or for a rendezvous and shelter in time of war." As for the hydrography, "I find that six fathoms may be carried through here and that the channel is perfectly feasible for battle ships. It could be closed with great ease and at short notice." In another report, he even went so far as to compare these waters to Nantucket and Vineyard sounds, which the Naval War College had studied in detail as a strategic exercise. It seemed "to hold the same relation to the Caribbean Sea that is held by the above mentioned Sounds to the waters of our North Atlantic coast," Knight observed. This was precisely the kind of report that Bradford and Dewey wanted. It demonstrates how hydrography and strategy mixed in the mind of a naval officer taught to think in this way by his training at the War College. "The General Board is impressed with the report of Lt. Cmdr. Knight," Dewey wrote to Bradford after the board reviewed it in June 1902, "and concurs in his estimate of the strategic value of the region in question." The board also recommended that Knight remain in command of *Yankton* for another surveying season as its strategic eyes on the water.[34]

It is impossible to say just how this new strategic view of the marine environment would have played out in a naval war—both world wars were too different in nature or distant in time to make meaningful connections—but some sense might be gleaned from the navy's winter maneuvers from

December 1902 to January 1903 off the coast of Puerto Rico and in the context of the Venezuelan Crisis, in which Germany and Great Britain blockaded and briefly bombarded Venezuela for unpaid debts, and thus called into question threats to the Monroe Doctrine. The old victor of Manila Bay, George Dewey, emerged from his seat at the head of the General Board to command a fleet "the largest ever assembled by the United States in time of peace," the press boasted. Its aims were practice, drill, and exercise, but President Theodore Roosevelt had also sought out Dewey for command with deteriorating conditions between Venezuela and its European creditors on his mind. Both men were Germanophobes, convinced that Germany intended to assert its own claims to the crumbling remnants of Spain's empire in the Caribbean and the Pacific. Officers at the Naval War College, meanwhile, had already embarked on a war plan that envisioned a clash with Germany in the Caribbean, in which an attacking force would seize an advanced base off the island of Culebra just east of Puerto Rico from which to launch an attack against the American fleet or its coast.[35]

The exercises were mundane enough in their daily routine, but the shadow of possible naval war in the Caribbean loomed. In the initial phase of maneuvers, the navy worked out a fleet problem in which an attacking force in mock battle successfully seized a port in Puerto Rico from another force arrayed to defend it. The fleet then engaged in a series of combined maneuvers based on scenarios worked out at the War College. All this seemed like McCarty Little's vision of war on the chart made real—naval officers taking their strategic understandings from the classroom and applying them to the waters of the American empire now in the shadow of real war. "Our Caribbean Sea naval maneuvers, with Admiral Dewey on the scene," reported the *Washington Evening Times* in October 1902, "promises to be as near like the real thing as the world's greatest living sea captain can make them."[36]

For Dewey and his chief of staff, Henry C. Taylor, the maneuvers were also an opportunity to assess firsthand the strategic features of American imperial waters, particularly those surrounding Culebra, where the exercises were taking place and which seemed sure to be a flash point in any Caribbean war with Germany. On January 3, Dewey and Taylor embarked on the yacht *Gloucester* for a steam around "The Sound" adjacent to Culebra, "a sheet of smooth water protected by an outlying line of reefs and islands," as Dewey described it. "All the staff was impressed by the value of this space of water," he recorded in his journal, "having sufficient depth for the largest battle ship and possessing space enough to anchor our entire fleet in absolutely smooth water—a quality of infinite advantage for coaling or for repairing damages in case of necessity." Dewey recommended a thorough survey and buoying of the four channels that permitted ample passage to and from the sound "so that in whatever direction an enemy may be approaching, our fleet can get out to attack him before he is aware of our presence." Taylor, in his official report, could hardly hide his enthusiasm. After noting soundings between eight and twelve fathoms, he declared its value "as a great naval base and rendezvous, and taken in connection with Target Bay roadstead, and the inner port of Great Harbor, as well as the remarkable bold hills, which protect these anchorages, offering every facility for gun defense, together with other qualities, make this vicinity of Culebra one of the most remarkable and valuable naval strongholds in the world." Here was Mahan's vision of a militarized sea made manifest: the navy's most influential postwar officers, strategic thinkers, and decision makers in awe over a body of water whose natural features made it an impregnable fortress and an ideal space from which to defend the empire from any attacker.[37]

Dewey had spent his days watching the exercises, noting in his journal the departure of his ships to be recoaled, and, most interesting, scribbling away on page after page cartographic representations of the harbors he had visited in the course of the exercises. Indeed, to anyone paging through the admiral's journal, the charts appear in extraordinary detail, the unmistakable product of many hours penned in leisurely contemplation of wars on the water that had dominated discussion at the War College and the General Board. Dewey was quite evidently having cartographic visions—no wonder. The General Board had spent the better part of its nearly two years of existence debating the environmental merits of so many bodies of water throughout the new empire. At sea once again, Dewey turned his mind, his eyes, and his pen to inscribe his own cartography of empire. There was his chart of Samaná Bay, Santo Domingo, and his chart of the *Gloucester*'s cruise around "The Sound" that had so captivated Dewey and his staff. Finally, there was the chart of Guantánamo Bay, where Dewey's steam yacht *Mayflower* called in January on its return to the United States. When *Mayflower* arrived, Dewey found the harbor strangely empty of ships and he noted flying fish jumping from the sea at *Mayflower*'s bow. Then he set to work on his own representation of the bay's contours, all the while overwriting on it the events of the day's cruise. The cartographic representations of the natural world thus formed the background for his daily activities during the winter maneuvers. These doodles evidently served no official purpose but to pass the time in strategic contemplation. Through it all—and quite liter-

Admiral George Dewey's sketch chart of Guantánamo Bay, January 1903, from the journal he kept while commanding the naval exercises off Culebra, Puerto Rico. Courtesy of the Library of Congress, Manuscript Division.

ally underscoring Dewey's journal—cartography, the environment, and the strategic waters of the new empire came to underlay the navy's daily work, an omnipresent vision of the waters it must defend and control in the new century.[38]

Summers at Newport became the center of naval strategic discourse as the Naval War College played out war games, constructed war charts, and heard lectures, while the General Board retreated there from the sweltering capital to take in the high social scene and discuss the most pressing matters facing the navy and the nation. Established in 1900 to advise the secretary of the navy, the General Board consisted of nine members from the various bureaus, the Office of Naval Intelligence, and the Naval War College. Admiral Dewey, the highest-ranking officer in the navy, presided. Between 1900 and 1903, the General Board's agenda largely revolved around naval bases and their location in the Caribbean and the Pacific.[39] Bradford at the Bureau of Equipment had so interposed himself in these discussions before the General Board that he became a member himself in August 1901 and stayed for more than two years, when, having pushed too strenuously for bases from the Philippines to the coast of China, he left the board. Almost from the start, then, the work of the General Board represented the confluence of these different, but interrelated forces within the navy— Bradford at the Bureau of Equipment; the surveyors and the hydrographer, Todd, under him; Captain Stockton and Rear Admiral French E. Chadwick as successive presidents of the Naval War College; and Dewey himself as the most influential officer in the navy. Officers at the college tested the principles gleaned from their games in Narragansett Bay or sometimes with the North Atlantic Fleet itself, while, between meetings, members of the General Board served as observers, umpires, and adjudicators. It should be no surprise that hydrography and the chart were foremost on their minds and important—sometimes determinative—tools of American empire.

All converged and mingled at Newport in strategic discussions about the hydrography of the American empire that amplified the place of charts and the marine environment in ways not before seen in the history of American naval science. In November 1901, Dewey reported to Secretary Long that Guantánamo Bay, due to its natural advantages, was the board's unanimous decision for an American naval base in the West Indies. From the beginning, the board had heard a string of arguments. Bradford had forwarded letters from Commander George L. Dyer and from Fletcher, each commanding yachts on the Cuba survey. He had also solicited and sent along opinions from commanders of the North Atlantic Fleet's battleships as well as from

Captain Bowman H. McCalla, who had led the attack on Guantánamo during the war. Guantánamo, Fletcher had reported, was "favorably situated" and "has a fine anchorage for a large fleet of vessels," while Dyer added that "Santiago is more easily defended than Guantánamo, but it is not so easy to enter." For his part, McCalla cited Guantánamo's defensibility and its spacious channel. "From a military point of view," and he perhaps understood this view more than anyone, "Guantanamo Bay would seem to be greatly superior to Santiago, for it is as easily defended, except as to mines, and its entrance is broad enough for four battleships to steam out in line." In August 1901, meanwhile, the board had reviewed the results of several strategic games at the War College, which, its members agreed, had "a bearing upon the solution of the Caribbean situation." Each game had pointed to the importance of Guantánamo as a base nearest the vital Windward Passage.[40]

On November 25, after considering all these points, the General Board went on record to the secretary of the navy, choosing Guantánamo chiefly for its geographical position and its environmental characteristics. "By reason of its position inside and adjacent to the Windward Passage, its commodious and well protected harbor and its easily defended entrance," Guantánamo, "commends itself strongly to the General Board as of the first importance for a naval base in Cuba," wrote Captain Robley D. Evans, summing up the board's opinion a few months later. By referring to the argument in favor of Guantánamo by the reflexive pronoun "itself," it was as if the bay's waters, which were broad at the mouth and well-protected within, had articulated their own argument—so powerful did these environmental features seem to suggest themselves to the observant officer and strategy maker. Here, again, was the agency of the environment seeming to assert its own claims to strategic advantage, from a natural enemy in the war with Spain to an ally capable of furthering American sea power and imperial defense more broadly, given more accurate knowledge and a deeper appreciation of its strategic assets. Guantánamo thus became the navy's outpost in the Caribbean, made official by a lease from the Cuban government to the United States in February 1903.[41]

The decision in favor of Subic Bay in the Philippines followed much the same course. Through the fall of 1901, a succession of officers had appeared before the General Board. There was Ellicott, who, in June 1900, presented the same lecture that he had given at the Naval War College on the importance of Iloilo and other points in the Philippines and the Pacific. Niblack, who had been a member of the Remey Board and had just made a survey of Subic, reported to the board in September 1901. Through the fall, the board discussed Remey's report, and the admiral himself appeared to answer questions in the spring of 1902. Perhaps more than anything, the weight of Admiral Dewey's opinion held sway. Any base in the Philippines had to be deep enough to accommodate the floating dry dock *Dewey*, named, of course, for the man who had done perhaps more than anyone to secure the Philippine Islands for the United States. So it was with intense personal interest that Dewey presided over and directed this debate. He was convinced that Subic was the superior position. "I may state from my own experience," he later wrote, referring to his victory at the Battle of Manila Bay in May 1898, that "I fully expected to find the Spanish fleet at Subig [*sic*] as from my strategical study of the situation that is where they should have been." By September 1901, the General Board had made up its mind. Subic, Dewey later reported in recapping the board's opinion, "possesses a capacious anchorage with sufficient depth of water for the largest ships, is capable of being excellently defended by fortifications and submarine mines," and "has good protection from the prevailing storms of the locality." The board, in its recommendation to Long, wrote that it was "impressed with the advantages possessed by Olongapo, in Subig Bay," and pressed him that "steps be taken toward the establishment of a strong naval base . . . with as little delay as possible." By the end of 1901, though many surveys and much work remained to be done across the empire, the navy had come to a conclusion regarding its positions in the Caribbean and the Far East. It had done so primarily based on hydrographic arguments and with reference to the charts and presentations of those who had completed the surveys themselves.[42]

In a curious turn, Bradford had dissented in both decisions. By 1901, for reasons that are not entirely clear, he had come to favor Havana in Cuba and Manila in the Philippines, and thus departed from the board's consensus. Perhaps he sensed in the future the ways in which the navy and the nation was beginning to back away from the fullest reaches of expansionist rhetoric and naval strategic planning in the years after 1898. With the decision made in favor of Guantánamo and Subic, congressional appropriations tightened and, at least in the Pacific, new logistical and strategic realities altered the navy's calculus. Ultimately, Subic never became the principal American naval base in the Pacific. By 1906, a series of debates within the navy and with the army had determined Subic to be too distant from the United States and indefensible from the growing naval power of Japan, which had won a decisive victory in the Russo-Japanese War in 1904–5. Manila, with its preexisting, albeit inadequate, infrastructure, ultimately served as the navy's base in the Philippines, while Pearl Harbor, five thousand miles to the east, became its fleet base in the Pacific. Whatever the reasons for Bradford's own conclusions in the naval base debate, he was quickly marginalized within the General Board. He continued to press the board and the secretary of the navy that matters relating to the establishment of coaling stations should reside primarily with him as chief of the Bureau of Equipment. It seems that Bradford, so attentive to the logistical importance of coal—now sought to protect and enlarge his influence in the bureau. In the summer of 1902, he had attempted unsuccessfully to assert himself in a dispute over rank as the senior member of the board next to Dewey. The secretary of the navy ultimately ruled in favor of Henry C. Taylor in this matter, whose experience as president of the War College made him a leading strategic voice within the navy. Bradford was not well liked. He was "a bit of a brute," remarked a fellow officer, though "with all his peculiarities . . . the most honest and single minded man in the Navy Department—no duty server." His unabashed expansionism to acquire coaling stations all over the Caribbean and the Pacific lacked the attentive focus of a strategic thinker attuned not only to the hydrography of imperial waters and the logistics of coal but also to the larger interplay among political, diplomatic, strategic, and hydrographic considerations required to defend the new empire.[43]

Indeed, it is important to note that hydrographic surveying and chartmaking, and the place of hydrographic discourse about the marine environment itself, operated within larger contexts that both furthered and limited its role in naval affairs. Its newfound significance to the navy, of course, was tied to the service's growing commitments around the world and Mahan's philosophy of sea power. But foreign countries, equating hydrography with strategic advantage, sometimes protested Americans charting their coastal waters. The new Republic of Cuba, for example, complained to the State Department about the presence of the navy's surveying ships—one small, yet illustrative example of the fraught place of Cuba's independence in the political and economic affairs of the island and within the larger American imperial presence in the Caribbean more broadly. Diplomatic complications were perhaps most marked in the navy's attempt to secure a base in the Chusan Islands. The Boxer Rebellion had proved that a base in the Philippines was not enough to support operations on the coast of China. Favorable in its hydrographic, strategic, and geographic features, the navy considered the Chusan Archipelago as a possible advanced base. But China could not sell the islands to the United States under a preexisting agreement with Great Britain. Indeed, the State Department worked closely with Dewey, Long, and Bradford in all matters related to prospective coaling stations in foreign waters. Hydrography was one important—if largely unacknowledged—factor within the larger political, diplomatic, economic, and cultural forces that swirled like eddies around the emergence of an oceanic empire both powerful in its vision, articulation, and practice and limited in its ultimate outcomes.[44]

Among the many factors that contributed to the fraught place of American power on the ocean at the turn of the twentieth century was, of course, the ocean. To American naval leaders, strategists, and surveyors it represented an old domain of American power reframed in new strategic and military imperatives. Attention to the marine environment, however, also illustrates just how frail and superficial this power could be. Impressive on paper and in the jingoistic rhetoric of American expansionism, the reality was significantly different. Focusing on naval science and the ocean environment can help us better see this imperial contradiction. As the congressional debate over appropriations in the ongoing bureaucratic and scientific rivalry between the Coast Survey and Hydrographic Office concluded that both could be usefully employed in surveying the waters of the new empire regardless of definitions of territoriality, there seemed, finally, a consensus among both agencies and the officials tasked with overseeing and running them that the hydrographic task at hand was immense. The natural qualities of the ocean—its depths, its vastness, its remoteness, its dynamism, and the maze of reefs, rocks, bars, and coasts where water met land—presented challenges to American cartography much bigger than any one agency's abilities to chart it. As the survey's superintendent, Henry S. Pritchett, told Congress in April 1900, "The resurvey of our long coastline . . . will never be finished," he concluded. "Is it not true," one congressman asked him, "that from the trenching of the ocean upon the land and the land upon the ocean, and the actions of currents, and the improvements of rivers and harbors, and variations in tides, and many other forces artificially or naturally require that this work should go on as long as those forces operate?" Pritchett responded, "Unquestionably."[45] Forces both natural and human were always making and remaking environments of commercial or strategic value. The state of hydrography was—quite literally by nature—always unfinished. As the setting and, indeed, an active agent in the course of maritime commerce and naval operations, the sea was an ever-changing environment that flouted the hydrographer and the nautical chart's attempts to keep pace with it. This work, as always, was an extraordinary undertaking for any two organizations, let alone one. The

Hydrographic Office and the Coast Survey thus proceeded, however ineffi-
ciently, to chart common waters and often to ask different scientific questions about its value to the navy, commerce, or marine science. Asked by a
member of the Naval Affairs Committee in 1902 whether there was "any
conflict now existing" between the two hydrographic institutions, Rear Admiral Bradford replied no. "There is plenty of work for both, and will be for
years to come?" the congressman pressed. "Yes," Bradford admitted.[46]

Nevertheless, by the turn of the twentieth century, as they considered the
merits of particular harbors and islands for military use, naval officers came
to believe that nature itself had imparted imperial waters with strategic
value. This appeared in a flurry of hydrographic activity along the coasts and
bays of the new American empire, producing charts for strategically important waters, but also transcending the traditional hydrographic chart to a
broader thematic cartography in the war games and war charts constructed
and debated at the Naval War College. Officers, for example, were consciously differentiating between commercial ports and those they termed
military, naval, or so-called man-of-war ports. Captain Charles D. Sigsbee, before the war a noted hydrographic technician and, after it, a member
of the General Board, wrote, "no harbor can now be considered a good man-
of-war harbor that will not admit a battleship fleet." For its part, the navy
needed only to survey, apprehend, and improve waters that seemed ready-
made for its battleships and the increasing logistical sprawl required to service, repair, and protect them. Thus, Mahan could write that Subic Bay was
"an impregnable fortress," which "lends itself most readily to defense." At
the Naval War College, Taylor and his students concluded that Long Island
was "endowed" by nature with "admirable strategic qualities." Still other
naval officers conducting surveys or reconnaissance reported to the department, citing the "magnificent natural advantages" of one harbor or another,
or the particular "facilities afforded by nature." In 1901, a General Board report on Samaná Bay, Santo Domingo, which the board deemed another flash
point in a possible war with Germany, cited an environment "fitted by nature for defense." Environmental questions thus became inseparable from
strategic considerations. The sea had military value, transformed in a vision
of Mahanian sea power whose meaning in the age of American empire was
inextricably related to the sea itself. The marine environment was omnipresent both beneath the keels of the navy's ships and in the minds of its commanding officers.[47] In this imagined cartography of empire, American sea
power emerged firmly fixed in the ocean environment itself with tremendous implications for the twentieth century and our own time.

# Epilogue
## Controlling the Great Common
. . . . . . . . . . . . . . . . . . . . . . . . . . . . . . . . . . . . . . . . . . . . . . . . . .

What manner of man is this that even the winds and the sea obey him?
—Book of Exodus

Stretch out thine hand over the sea and divide it.
—Book of Matthew[1]

There is a story—perhaps partly apocryphal, but probably not—about the
writer John Steinbeck and his friend, the marine biologist Ed Ricketts. As
Steinbeck wrote in his memorial following Ricketts's death, the two had just
returned from a scientific research trip to the Gulf of California that formed
the narrative for Steinbeck's *The Log from the Sea of Cortez*. It was early 1942.
The Japanese had attacked Pearl Harbor, and the United States was at war.
In preparing their research findings and attempting to put their work in the
context of the broader marine biology of the Pacific Ocean, Steinbeck and
Ricketts had come across a number of scientific monographs written by
Japanese scientists. Steinbeck surmised that they had been commissioned
by Tokyo to study the islands administered by Japan as League of Nations
mandates after Germany lost World War I. As "good scientists and special-
ists," these Japanese had published their secret research as academic studies available, presumably, to their "friends all over the world who would
appreciate and applaud their work in pure science." To Steinbeck and Ricketts's astonishment, they had stumbled on hydrographic and zoological
studies of the littoral waters of Japanese-held Pacific islands, which were
now squarely in the crosshairs of the American military's amphibious coun-
teroffensive across the Central Pacific. What followed, as Steinbeck recalled, was "truly comic opera."[2]

Aware of the probable import of their discovery, Steinbeck wrote Secretary of the Navy Frank Knox. "It is not generally known," Steinbeck told
Knox, "that the most complete topographical as well as faunal information
about any given area is found in the zoological and ecological reports of
scientists investigating the region." Soon thereafter, a lieutenant commander

from the Office of Naval Intelligence appeared at Ricketts's Pacific Biological Laboratories along Monterey's Cannery Row to investigate. Presented with the monographs, the naval officer was skeptical. "Do you speak or read Japanese?" the officer asked Ricketts. "No," he replied. "Does your partner speak or read Japanese?" Again, he answered no. "Only then," Steinbeck wrote, "did Ed understand him." The naval officer could hardly believe that such studies would be written in English. This, of course, was the international language of science, and so it was only natural that these Japanese scientists wrote in English for a profession that transcended national borders and military-strategic contests of power. "This thought, Ed said, really made quite a struggle to get in, but it failed." The officer promised, "you will hear from us." But they never did. As Steinbeck quipped, "I have always wondered whether they had the information or got it. I wonder whether some of the soldiers whose landing craft grounded a quarter of a mile from the beach and who had to wade ashore under fire had the feeling that bottom and tidal range either were not known or ignored. I don't know. Thus was our impertinent attempt to change the techniques of warfare put in its place. But we won." Perhaps Steinbeck intended to say that the United States won the Pacific War without accurately knowing the hydrography of so many Pacific islands. Or perhaps he had meant to say that he and Ricketts had been validated when the Marines met folly, and near disaster, on the coral reefs of Tarawa atoll. The United States took the atoll from the Japanese in November 1943 with a loss of more than one thousand American dead, many of whom never reached the beach. In planning the amphibious assault, the Americans had used a chart constructed by Charles Wilkes and the Ex. Ex. in 1841, nearly a century earlier. "Ignorance of how to tackle a strongly defended coral atoll surrounded by a fringing reef," argued historian Samuel Eliot Morison in the navy's official history, "was responsible for most of the errors." Perhaps, also, the tide and the reefs themselves had won, humbling one of the most powerful invasion forces the world had ever seen. Whatever the case, Steinbeck and Ricketts's brief encounter with Naval Intelligence speaks to the continued—indeed, the expanded—importance of hydrography and the sea in the context of twentieth-century warfare, the ongoing challenges that the navy experienced in understanding and thinking about the marine environment, and the continued divide between the military and civilian worlds, particularly where science was concerned.[3]

The sea both made American empire and unmade it, and in the years after the turn of the twentieth century, the navy's relationship with marine science, the marine environment, and hydrographic surveying took twists and turns, rode crests and submarine valleys largely driven by the exigencies of twentieth-century war. World War I and World War II opened new undersea dimensions for the navy in mine warfare, amphibious assaults, and, most important, submarines. As historian Gary Weir has shown, this prefaced a rapprochement with the civilian marine sciences based in common interests and a mutual need both for academically trained oceanographers and for the vessels and wartime funding the navy could provide. In the interwar years, civilian scientists and the navy forged a sometimes awkward, but largely cooperative, relationship, which only grew during the postwar era in the context of a Cold War that privileged science as a tool of national security, defense, and war making. Science itself could be weaponized in the military struggle for victory and the concurrent ideological struggle for minds, which might be achieved, as Jacob Darwin Hamblin has argued, through strategic alteration to the natural world itself. One need only conjure the giant half-mile-wide column of water swallowing a ghost fleet of World War II relics during the 1946 atomic Baker Test at Bikini Atoll, Marshall Islands, to grasp something of the relationship among science, the marine environment, and the extraordinary growth of the American military's destructive power. The postwar national security state spawned new institutions such as the Defense Mapping Agency, which consolidated the various cartographic organizations within the Department of Defense and constructed a Cold War cartography that privileged mapping as a form of intelligence, and thus strategic and geopolitical power. Apart from the specter of atomic warfare, in Vietnam the navy found itself embroiled not in great Mahanian sea fights, but in the so-called brown water patrol and blockade work of Operation Game Warden and Market Time along the circuitous and always-changing waterways of the Mekong Delta, an environment in which the navy was not adequately prepared to fight but nevertheless was reminiscent—broadly speaking—of littoral operations from the Mexican War to the War with Spain.[4]

Nevertheless, reminding the navy that the sea mattered became something of an annual lament among hydrographers within an institution that did not always recognize the value of naval science and did not always have leaders like Maury or Bradford, or conflicts like the war with Spain to keep in the forefront of the navy's peacetime work. In a 1970 article published in the United States Naval Institute's *Proceedings*, Charles C. Bates, George Tselepis, and Daniel von Nieda, all hydrographers, called for "shallow thinking. Where and when today's enemy cannot exploit the environment of blue water and seeks sanctuary in the brown," they argued, "the U.S. Navy must range the shallows in craft which, drawing only four feet of water, can get by without a paddle—but not without a chart." This should have been a familiar argument, but one not always heeded in a navy whose historic relationship with hydrography and marine science had often been conflicted or complicated. As the powerful chief of the Bureau of Navigation Arent S. Crowninshield put it early in the new century, naval scientists were not experienced enough "in ordnance, or in the handling of vessels . . . and men," in other words, the fundamental qualities of the naval officer's profession. Each and every officer," he declared, should "find a place aboard battleships, training vessels, cruisers, or torpedo boats." As early as 1905, hydrographers such as Lieutenant Commander Harry M. Hodges once again took up the cause of naval science, attempting to call attention to the importance of scientific work within the ranks of a profession that did not often equate naval scientists with sea warriors. "I wish to repeat," Hodges reminded the Navy Department, "that the work of this Office and of its various branches consists in giving the greatest possible amount of assistance to the Navy and merchant marine. Also," he continued, "I am still of the opinion that it is not generally realized in the Department how wide a scope the functions of the Hydrographic Office take." The navy's embrace of hydrography—so central to the transformation of empire during the nineteenth century—was nevertheless fleeting. Punctuated by staccato-like moments of intense hydrographic activity, by grand exploring expeditions, or the visions of a Maury or a Bradford, naval scientists advanced American interests over the sea at particularly momentous times of American commercial and strategic necessity. These, however, were short, albeit transformational bursts, and they continued into the twentieth century—a testament both to the important needs for accurate information about the marine environment and, at the same time, the navy's conflicted views about the utility and place of science within its ranks.[5]

While the Navy never quite embraced in any long-term, sustained way the newfound strategic relationship forged at the turn of the century between hydrography and the marine environment it sought to control, the Marine Corps did, largely in the person of Lieutenant Commander Earl H. Ellis, whose clandestine mission to the Japanese-mandated islands of the Western Pacific from 1921 until his mysterious death in 1923 led to the influential *Advance Base Operations Micronesia* and the Marines's doctrine of amphibious assault against heavily defended island beaches. Ellis's work transformed the Marine Corps and prefaced the trans-Pacific offensive of 1942–45 codified in the Navy's War Plan Orange, its strategic contingency for war against Japan. In formulating the amphibious assault, Ellis's eyes remained solidly fixed, among other things, on the beach and the marine environment. "Perhaps there is no other area in the world where navigation is more difficult," Ellis wrote, referring to the Western Pacific. "It has never been accurately charted," he added. "Oftentimes navigation is dependent entirely upon the eye . . . islets and reefs generally rise abruptly from great depths" and "the ocean currents to be met with are irregular and uncertain." Where did Ellis's strategic vision of the natural environment originate? From 1911 to 1913, he served, first, as a student and then as an instructor at the Naval War College. In a lecture he gave to students and staff in 1913, Ellis argued that the selection of bases, for example, should have "a good anchorage for at least half the force likely to be based thereon, protection from the elements, an entrance admitting of easy entrance and egress, healthy climate," and "a small range of tide and weak tidal currents." Ellis' doctrinal insights, which altered the course of the Marine Corps in the interwar years, lay in his course of study in Newport where he learned to see the marine environment strategically.[6]

Like previous centuries, the twenty-first poses new challenges to the world's most powerful sea service, again calling on naval science—present and past—to meet threats to the United States. Among them is that posed by the marine environment and the environmental changes now occurring as a result of climate change and other human-produced alterations to the natural world. In 2010, the Department of Defense recognized that climate change "may act as an accelerant of instability or conflict" and noted "the impacts of climate change on our facilities and military capabilities." The sea is now changing in ways that threaten the national security of the United States. The Navy must grapple with a new set of strategic challenges that are related to—but also quite different from—broader scientific and political debates about environmental change. A projected increase in sea levels between one and four feet over the next century not only threatens coastal communities and ecosystems, but also could submerge parts of Norfolk, Virginia—site of the world's largest naval base. Elsewhere, ice-choked Arctic sea lanes begin to open and with them the question of who must guard them and who may claim the resources below. Droughts and other natural disasters increase competition and conflict for resources, land, and sovereignty. Low-lying island nations are slowly sinking beneath the ocean. The People's Republic of China builds artificial islands atop half-submerged reefs in the South China Sea in defiance of international maritime laws. In many ways, the sea has become a barometer for the ways that humans have changed the natural world to suit political, economic, and strategic needs.[7]

As the sea itself threatens national security, the Navy has struggled to articulate the ways it now accounts for environmental change within a larger strategic discourse that has long focused on human foes, nation states, and their militaries and navies. In a 2016 radio interview with WNYC, retired rear admiral David Titley, former oceanographer of the Navy, former chief operating officer of the National Oceanic and Atmospheric Administration, and former director of the Navy's Task Force Climate Change—a leading voice in the debate over the strategic implications of environmental issues—observed that "as a military issue," climate change presents "challenges and even threats to U.S. security." He went on to clarify that he believed the climate is "a risk more than a threat," referring to Soviet nuclear brinksmanship during the Cold War as a historical example of a threat whose definition rested on intention that was implicitly human. "The climate is a little different," he added, "it doesn't have intentions, it's just physics. The ice doesn't care who wins this next election; the ice doesn't care who controls Congress; the ice just melts." How, Titley seems to wonder, will the Department of Defense deal with and even talk about risks or threats to national security that are nonhuman?

While the navy's Oceanographic Office and a special Task Force Climate Change directed by the oceanographer of the navy have brought renewed attention to environmental change at sea both within the service and to a broader public grappling with the larger questions of humans' relationship to the natural world, climate scientists are beginning to look to the past for a firmer idea of just how much the world's oceans have changed over the last two centuries. Oldweather.org, a crowd-sourced digital history project sponsored in part by the National Oceanic and Atmospheric Administration, the U.S. Coast Guard, and the National Archives and Records Administration, is now seeking the public's participation in combing through historic ship logs like the kind that Matthew Fontaine Maury was so interested in collecting and reviewing to understand the ways the world's climate has changed over time and what that may mean for the future of the planet. In ways reminiscent of Maury's antebellum system of research, these long-gone mariner observers have once again found life—their logs, newfound utility—in an era of tremendous uncertainty about a changing natural world. Such initiatives suggest how new work in climate science and oceanography are beginning to use old sources to examine the present and future state not only of the world's oceans but of the natural world itself. The seeds for understanding the present and future, in other words, lay in our historical relationship with the ocean.

As I have argued here, the navy's historical encounter with the marine environment is an important part of that story and of the larger extension of American empire over the ocean. It is certainly not the only part, nor is the commercial and strategic role of hydrographic surveying and chartmaking the only corner of naval science or marine science to shed light on these historical questions. Yet, the history of the navy's hydrographic efforts in the nineteenth century throughout a burgeoning oceanic empire sheds light on questions of power that are central to understanding the growth of the United States, its imperial ambitions, and the relationship between humans and a watery natural world that often revolved—in its many definitions—around notions of control. We are still grappling with our ability (or inability) to control a marine environment that increasingly threatens not just American national security but our coastal cities and a larger natural world in the midst of climatic changes that bring drought, flood, and fire in increasingly erratic and exaggerated ways to the United States and the world. For the navy and the nation more broadly this has meant preserving and protecting an ocean environment that seems both increasingly fragile and increasingly threatening as the grounding of the USS *Guardian*, with which this book began, illustrates. Once the bane of navigators—the perilous navigational danger looming beneath uncharted seas—the reefs themselves now face destruction, not only by modern vessels but also by a host of human-induced environmental factors, from rising sea temperatures to coastal erosion and ocean acidification.[8]

The sea has long been and continues to be a force with which the navy must reckon. Its efforts in surveying and charting a vast, deep, and dynamic marine environment both hastened the emergence of an American empire and constantly reminded imperialists—from surveyors to naval commanders afloat and strategic decision makers—of the inherent limits of their understanding and the tremendous challenges of conducting scientific work at sea. Despite nearly two centuries of institutionalized surveying and charting, the seventh-tenths of the world covered by salt water and the increasingly sophisticated set of technologies employed to do so, oceanographer Sylvia Earle reminds us that we still have only charted about 2 percent of the world's oceans in any comprehensive way. Much of the deep, as she and others have pointed out, remains more mysterious to us than the surface of the moon. As the United States expanded first its commercial aspirations and later its strategic vision over the world's oceans during the nineteenth century, the hydrographic process of surveying and chartmaking reflected the pretension of naval science that the natural world could be known, ordered, and controlled—made to serve the expanding interests of the United States. Ultimately, for the navy, sea power rested fundamentally in knowledge of the sea itself. Yet such efforts to control the marine environment also exposed the follies, contradictions, and—ultimately—the impossibility of completely doing so. In 1890, the navalist, imperialist, and theorist Alfred Thayer Mahan wrote of "controlling the great common." My hope is that this book—aside from what Mahan actually meant by that line—suggests both the navy's long-standing intention to control the sea and its inability to fully fathom a great common of competing and contesting forces—political, military, commercial, scientific, indigenous, and, of course, environmental.

# Acknowledgments

"Exultation is the going of an inland soul to sea," the poet Emily Dickinson once wrote. "Bred as we, among the mountains, can the sailor understand the divine intoxication of the first league out from land?" I admit that I am mostly a landlubber, happy to be near saltwater, but convinced, as the American writer Richard Henry Dana suggested, that seafaring soon disabuses the inland soul of what he called "the witchery in the sea," or the divine intoxication to which Dickinson refers. I am more comfortable in the classroom both as a student and teacher, in the archives, or clumsily hammering away on my keyboard—half cat-crazed—early in the morning and late into night. Nevertheless, the sea has been constantly with me in my mind. There is a part of me that imagines I have been to sea for much of the last decade—a passenger if not, thankfully, the navigator, shipmaster, or, God forbid, the surveyor—by virtue only of reading so many charts, journals, correspondence, and other texts, and I am immeasurably grateful to countless people who helped me to find these sources, interpret them, and make sense of them. Without so many personal, intellectual, and logistical debts, this book would not exist. I am happy to acknowledge and thank them now at the end of this voyage. Of course, any errors of fact or interpretation are mine alone.

In many ways, the long evolution of this project mirrored my own intellectual growth. Ideally, graduate education should work that way, and for that, I am grateful to the members of my dissertation committee at Temple University, each one of whom changed the way I thought and wrote about the past in their courses and, later, by their counsel. Aside from my dad and Mrs. Rice—my high school AP English teacher—Gregory J. W. Urwin has had the most influence on my growth as a writer. If you look closely, you will see his incisive influence present all over the margins and between the lines of this book. My advisor, mentor, and an accomplished military historian, Greg encouraged me to move outside the traditional bounds of the subfield. I remember submitting chapter after chapter on surveys and charts, and then receiving his edited copy of the dissertation's penultimate chapter on the Spanish-American War with the comment, always in red pen, "finally, a war!" He is an untiring champion for his students. Beth Bailey's influence as a historian of war and society and cultural history pressed me to think about maps and environments as cultural, intellectual, and imagined constructs. To Beth goes the credit for mentioning that what I was really trying to tell was a story about American empire. Without her guidance, this book would be something much less, a litany perhaps of surveys and charts without nearly as much of the "so what?" Finally, Drew Isenberg's environmental history course was a watershed, and to him goes the credit for pressing a military historian to more deeply and fully see the relationship between humans and the natural world. To the extent that this is an environmental history of naval affairs and American empire, Drew's influence is inextricable. At Temple, I also wish to thank Howard Spodek and Jay Lockenour, both generous and forgiving teachers, as well as Todd Shepard for guiding this project in the spring of 2006 where it began as a research seminar paper.

Graduate student comradery is a special kind of friendship, indeed, and I was fortunate to learn, celebrate, and commiserate with a bunch of smart, good people. From beginning to end, Mike Dolski showed me how to be a historian. I am certain that I would never have gotten far along without him. We survived, Mike, surely despite ourselves. Thanks! I'm grateful to Josh Wolf, in so many ways a civil warrior and now an avid scholar of the Old Sail Navy. Josh has impeccable taste in music, and, besides, he lingered with me one night in Lexington, Virginia, to sing "Thunder Road." I'll never forget that, nor our emergency run to Peebles for a conference suit woefully short in the legs. There are too many good stories to tell in this way, but nearly all of them involve Mike, Josh, and Eric Klinek, Jean-Pierre Beugoms, Marty Clemis, E. J. Catagnus, and Rich Grippaldi. Thanks also to Dave Ulbrich, Mike Lynch, Chris Golding, Bob Diehl, and Temple alums Jennie and Pat Speelman for inviting me for one wonderful trimester to the United States Merchant Marine Academy.

A second source of support along the way has been the United States Navy. I am particularly grateful for the guidance of John B. Hattendorf, Ernest J. King Professor of Maritime History at the United States Naval War College. John is the eminent scholar of American naval and maritime history, and is as generous, humble, and unassuming as any whom I have met. I am honored to have had him as an outside reader of my dissertation. I am also grateful for the support of the Naval History and Heritage Command (NHHC) through its Rear Admiral John D. Hayes Pre-Dissertation Fellowship and the Naval War College through its Edward M. Miller Fellowship. At NHHC, I am grateful to Michael Crawford, Curtis Utz, Bob Cressman, and Ed Furgol, and at the Naval War College to Evelyn Cherpak and David Kohnen. At the Bureau of Medicine, I am grateful to Jan Herman for giving me a tour of the old Naval Observatory. Finally, thanks to the History Department, the midshipmen, and the Class of 1957 at the United States Naval Academy (USNA), where I spent two memorable years as the Class of 1957 Postdoctoral Fellow in Naval History. It was an honor that I will always cherish. At USNA, thanks, in particular, to Captain Bill Peerenboom, USN, Ret., and his classmates, to C. C. Felker, Davin O'Hora, Fred Harrod, Bob Love, Nancy Ellenberger, Aaron O'Connell, Tom McCarthy, Rick Ruth, Richard Abels, Lori Bogle, Mark Belson, Sharika Crawford, Tom Brennan, Mary DeCredico, Chris Rentfrow, Kelcy Sagstetter, Joe Slaughter, Ernie Tucker, Miles Yu, Don Wallace, Larry Burke, Sean Getway, and B. J. Armstrong. As a postdoc fellow, my work was guided by lunches, car rides, and office gams with two distinguished visiting naval historians, Bill Trimble and Dave Rosenberg.

The community of maritime historians has been equally generous to me. I am indebted to Mystic Seaport Museum and its Paul Cuffe Memorial Fellowship, its Frank C. Munson Institute of American Maritime Studies and codirectors Glenn Gordinier and Eric Roorda, and to the faculty, including Helen Rozwadowski and Rich King, and, finally, the Seaport for the opportunity to participate in the 38th Voyage of the museum's historic whaleship the *Charles W. Morgan* in the summer of 2014. At Mystic Seaport Museum's Collections Research Center, Paul O'Pecko, Maribeth Bielinski, and Carole Mowrey have been generous with their time and knowledge.

I am grateful to a number of editors for seeing promise in my scholarship, sending it to peer reviewers, publishing it, and for permission to reproduce some of that work herein. Thanks to Nancy Langston and Bruce Vandervort, and a small army of anonymous peer reviewers over the years, who always pressed me to think longer, harder, and deeper about my interpretations and assumptions. Thanks especially to the University of North Carolina Press and Brandon Proia, who offered sage and timely advice, and expressed faith in this project, even as it existed in a very primitive state.

The world of archives remains confusing to me, but not nearly so now as when I started this project. For that, I owe a tremendous debt to the many archivists, research institutions, and libraries not yet mentioned above, including the Library of Congress and, in particular, Michael Klein and Cynthia Smith, the National Archives and Records Administration, especially Mark Mollan, the Beinecke Rare Book and Manuscript Library at Yale University, the Archives and Special Collections at Franklin & Marshall College, the Special Collections at East Carolina University, especially Dale Sauter and Tyler Caldwell, the New Bedford Whaling Museum, the staff of the Navy Department Library, as well as the libraries of Temple University, Hood College, the U.S. Naval Academy, the Sterling Memorial Library at Yale University, and the Buley Library at Southern Connecticut State University.

The latter has become my new home, and I am grateful for the support and guidance of my colleagues in the History Department, especially Troy Paddock, Steve Amerman, Darcy Kern, Ginny Metaxas, Byron Nakamura, Tom Radice, Michele Thompson, Polly Beals, and Steve Judd, all of whom read a version of my much-beleaguered, endlessly revised introduction during a departmental reading group. I am so grateful to join them, and to teach history in higher education with some degree of permanence and job security.

My first taste of academic history came as an undergraduate at West Chester University of Pennsylvania, where Kevin Dean, director of the Honor's College, created an environment that helped me grow as a person and student. In the History Department, I was fortunate enough to unwittingly stumble on a faculty that boasted two naval and maritime historians. Thomas J. Heston, my undergraduate advisor, convinced me that teaching history in college was what I wanted to do. One day in the middle of a lecture, he mentioned that if he could write a dissertation again, he might consider the scientific work of Matthew Fontaine Maury in the context of

American empire at sea. I took note. Finally, at West Chester, my mentor and good friend Tom Legg ushered me through the process of graduate school applications with sober realism, and he has been there to share the joys and sorrows of every step of my journey. He is the best teacher I know in and out of the classroom.

Finally, to my family, who have nurtured and indulged this project and my love for history since childhood. Thanks to my mom for her endless curiosity and for always pressing me out of familiar, comfortable things. Thanks to my dad, expert grammarian, who read so many of these pages in their various forms and found many a missing comma and not a few split infinitives. To my grandfathers, in particular, both veterans, I owe my fascination for military history. Thanks to my extended family, Pam, Ron, and Kara. Finally, I am forever grateful to Megan and Nel for bearing so many absentminded days, weeks, and months in an effort to bring this project to a close. Megan has been my partner in this as in everything. Her commitment to it, her encouragement, and her love are present in every page. To Nel, who appeared just as the difficult job of turning a dissertation into a book seemed most pressing, I am grateful for bringing perspective to my work and clarity to my life. She loves the ocean as much as her mother and I do. When she looks out to sea, she reminds me to look forward, not back.

The nineteenth-century mariner Robert Weir, with whom this book begins, was looking forward when his whaleship, the *Clara Bell*, finally reached New Bedford, Massachusetts, in May 1858 after three long years at sea. "We touched the wharf and I touched the shore—pure bona fide American land hurrah!" I feel something of Weir's relief and exhilaration at the close of this, my own kind of voyage.[1]

# Notes

## Introduction

1. See Floyd Whaley, "U.S. Navy to Scrap Vessel Stuck on Philippine Reef," *New York Times*, January 31, 2013.

2. Alistair Sponsel, "Coral Reefs as Objects of Scientific Study, from Threatening to Threatened" (paper presented at the Annual Meeting of the History of Science Society, Chicago, November 2014).

3. Mahan, "Discussion of the Elements of Sea Power," in *The Influence of Sea Power*, 25–89; thanks to an anonymous reader for clearly pointing out the nuts and bolts level of this project's significance.

4. Steinberg, *The Social Construction of the Ocean*, 89–109.

5. On the state of U.S. naval history, see Sumida and Rosenberg, "Machines, Men, Manufacturing, Management and Money," in Hattendorf, *Doing Naval History*, 25–40; see also Hattendorf, *Ubi Sumus?*

6. See Pinsel, *150 Years of Service on the Seas*; Dick, *Sky and Ocean Joined*; Bates, *HYDRO to NAVOCEANO*; Ponko, *Ships, Seas, and Scientists*; Jampoler, *Sailors in the Holy Land*; Philbrick, *Sea of Glory*; and Williams, *Matthew Fontaine Maury*.

7. Reidy, *Tides of History*, 273, 294; Ritchie, *The Admiralty Chart*, 189.

8. James K. Paulding to Charles Wilkes, Navy Department, August 11, 1838, in Charles Wilkes, *Narrative of the United States Exploring Expedition during the Years 1838, 1839, 1840, 1841, and 1842*, vol. 1 (Philadelphia: Lea and Blanchard, 1845), xxviii–xxix.

9. On American antebellum empire, see, for example, Schroeder, *Shaping a Maritime Empire*; Greenberg, *Manifest Manhood*; and Igler, *The Great Ocean*; see also Goetzmann, among others, *Exploration and Empire*. Two recent works considering the antebellum roots of American commercial empire at sea are Morrison, *True Yankees*, and Rouleau, *With Sails Whitening Every Sea*.

10. Schlee, *The Edge of an Unfamiliar World*, 63.

11. See Theberge, *The Coast Survey, 1807–1867*, vol. 1; Slotten, *Patronage, Practice, and the Culture of American Science*.

12. Harley and Woodward, *The History of Cartography*, 1:1.

13. Akerman, *The Imperial Map*; Jacob, *The Sovereign Map*, 23, 67; Harley and Woodward, *The History of Cartography*, 1:xv–xxi; Burnett, *Masters of All They Surveyed*; Edney, *Mapping an Empire*; Winichakul, *Siam Mapped*; Petto, *Mapping and Charting in Early Modern England and France*; Schulten, *Mapping the Nation*; Anderson, *Imagined Communities*, 170–78; Said, *Orientalism*.

14. See Brückner, *The Geographic Revolution in Early America*; Short, *Cartographic Encounters*; and Schulten, *Mapping the Nation*.

15. Three exceptions are Reidy, *Tides of History*, and Burnett, "Hydrographic Discipline among the Navigators," in Akerman, *The Imperial Map*, 185–259; Petto, *Mapping and Charting in Early Modern England and France*.

16. See Nash, "The Agency of Nature or the Nature of Agency," 67–69, and Steinberg, "Down to Earth," 798–820.

17. McNeill, *Mosquito Empires*, 2.

18. Private Journal of William Reynolds, October 8, 1839, 1:237, transcribed by Thomas Philbrick, Reynolds Family Papers, Archives and Special Collections, Franklin and Marshall College, Lancaster, PA.

19. United States Navy, *Annual Report of the Secretary of the Navy for the Year 1903*, 365.

20. See Tucker and Russell, *Natural Enemy, Natural Ally*; Russell, *War and Nature*; Drake, *The Blue, the Gray, and the Green*; Nelson, *Ruin Nation*; Brady, *War upon the Land*; and Keller, "The Mountains Roar," 254–74.

21. Steinberg, *The Social Construction of the Ocean*, 207; see Cronon, "The Trouble with Wilderness; or, Getting Back to the Wrong Nature," in *Uncommon Ground*; Lewis, *American Wilderness*; Nash, *Wilderness and the American Mind*. On wilderness in the context of American military engineering and war making, see Brady, *War upon the Land*; in a maritime context, see Kroll, *America's Ocean Wilderness*.

22. Rozwadowski, *Fathoming the Ocean*, 40.

23. Bolster, "Opportunities in Marine Environmental History," 575; see also Gillis and Torma, *Fluid Frontiers*.

24. See Bolster, *The Mortal Sea*; Rozwadowski, *Fathoming the Ocean*; McKenzie, *Clearing the Coastline*; Keiner, *The Oyster Question*; and Pastore, *Between Land and Sea*.

25. See Lipman, *The Saltwater Frontier*; Nash, *The Urban Crucible*; Gilje, *Liberty on the Waterfront*; Vickers, *Farmers and Fishermen*; Perl-Rosenthal, *Citizen Sailors*; Bolster, *Black Jacks*; and Rediker, *The Amistad Rebellion*.


## Chapter One

1. Diary of Robert Weir, *Clara Bell*, Log 164, April 12, 1858, Collections Research Center (CRC), Mystic Seaport Museum (MSM), Mystic, CT.

2. Vickers with Walsh, *Young Men and the Sea*, 149, 198; McKee, *A Gentlemanly and Honorable Profession*, 400–402.

3. Diary of Humphrey Hill, *Magdala*, Log 908, September 4, 1849, CRC, MSM.

4. Dana, *Two Years before the Mast*, 347; Weir diary, April 5, 1858, CRC, MSM; Diary of Nelson Cole Haley, *Charles W. Morgan*, Log 145, CRC, MSM, 330.

5. Rediker, *Between the Devil and the Deep Blue Sea*, 185–86; White, *The Organic Machine*, 3–29.

6. On maritime folklore, see Beck, *Folklore and the Sea*, 90–91, 118, and Baker, *Folklore of the Sea*.

7. Cooper, *The Red Rover*, 203; Dana, *Two Years before the Mast*, 38–39. On the difficulties of scientific practice at sea, see Rozwadowski, *Fathoming the Ocean*, 177–209; Burnett, *Trying Leviathan*, 209; see also Brown, "A Natural History of the Gloucester Sea Serpent," 402–36.

8. Melville, *Moby-Dick*, 861. See also Greenberg, *Manifest Manhood*; Lyons, *American Pacificism*; Morrison, *True Yankees*; and Rouleau, *With Sails Whitening Every Sea*.

9. Philbrick, *James Fenimore Cooper*, 71; Melville, *Moby-Dick*, 910; Philbrick, *James Fenimore Cooper*, 102, 192.

10. Diary of William Abbe, *Atkins Adams*, 189, Log 485, New Bedford Whaling Museum Research Library and Archives, New Bedford, MA.

11. Morrison, *True Yankees*, xvii–xviii; Rouleau, *With Sails Whitening Every Sea*, 3; de Tocqueville, *Democracy in America*, 427. See also Roland, Bolster, and Keyssar, *The Way of the Ship*, 429; Creighton, *Rites and Passages*, 36; Dolan, *Leviathan*, 206; and Rozwadowski, *Fathoming the Ocean*, 47.

12. Schroeder, *Shaping a Maritime Empire*, 3; Morison, *"Old Bruin*," 326–29; Hagan, *This People's Navy*, 149.

13. Blunt, *The American Coast Pilot*, ill; Diary of Unknown Author, *Weymouth* and *Indramayoe*, Log 507, July 20–22, 1823, CRC, MSM.

14. See Dodge, "Fiji Trader," 3–19; Philbrick, *Sea of Glory*, 194.

15. Endicott, *Wrecked among Cannibals in the Fijis*, 20; Oliver, *Wreck of the Glide*, 27. See also Ledyard, *The Last Voyage of Captain Cook*.

16. Wallis, *Life in Feejee*, 94; *Salem Gazette*, June 2, 1835, quoted in Ward, *American Activities in the Central Pacific, 1790–1870*, 2:385.

17. Blum, *The View from the Masthead*, 109–10; Cochelet, *Narrative of the Shipwreck of the Sophia*; Riley, *Authentic Narrative of the Loss of the Brig Commerce*, 9, quoted in Blum, *View from the Masthead*, 66. On the intersection of American literature, science, and empire, see Walls, *The Passage to Cosmos*.

18. Cooper, *History of the Navy*, 78

19. Porter, Journal of a Cruise Made to the Pacific Ocean, 1:v.

20. Ibid., 88–89.

21. Porter, *Journal of a Cruise Made to the Pacific Ocean*, 1:72, 88–89.

22. Schroeder, *Shaping a Maritime Empire*, 116; Smith, "The Navy before Darwinism," 41–55; Pinsel, *150 Years of Service on the Seas*, 1:1–10; Dick, *Sky and Ocean Joined*, 17–44; Bates, *HYDRO to NAVOCEANO*, 1–3; Weber, *The Hydrographic Office*, 1–17; Hughes, *Founding and Development of the U.S. Hydrographic Office*, 1–17.

23. Campbell, "Portolan Charts from the Late Thirteenth Century to 1500," in Harley and Woodward, *Cartography in Prehistoric, Ancient, and Medieval Europe and the Mediterranean*, 1:372; see also Taylor, *The Haven-Finding Art*, 35–64; Harley and Woodward, *Cartography in Prehistoric, Ancient, and Medieval Europe and the Mediterranean*, vol. 1 of *The History of Cartography*, 148–200.

24. Reidy, *Tides of History*, 6; Ritchie, *The Admiralty Chart*, 5, 14–15. On hydrography in the Royal Navy, see also Hornsby, *Surveyors of Empire*, and Morris, *Charts and Surveys in Peace and War*. On the pursuit of correct longitude, central to the science of marine navigation, see Andrewes, *The Quest for Longitude*; Howse, *Greenwich Time and the Discovery of the Longitude*; and Rozwadowski, *Fathoming the Ocean*, 48.

25. McKenzie, "Vocational Science and the Politics of Independence: The Boston Marine Society, 1754–1812." On Bowditch and Blunt, see Burstyn, *At the Sign of the Quadrant*; Burstyn, "Seafaring and the Emergence of American Science," in *The Atlantic World of Robert G. Albion*, 76–80; and Reingold, *Science in Nineteenth-Century America*, 11–14.

26. Ferguson, *Truxtun of the Constellation*, 56; Blum, *View from the Masthead*, 36; H. B., "Scraps from the Lucky Bag, No. III," *Southern Literary Messenger* 6 (December 1840): 790.

27. On navigation and educational reform in the navy, see Leeman, *The Long Road to Annapolis*, 146–57; on the professional aspirations of the nineteenth-century American naval officer corps, see McKee, *Gentlemanly and Honorable Profession*, and Karsten, *The Naval Aristocracy*.

28. United States Navy, *Rules, Regulations, and Instructions, for the Naval Service of the United States* (Washington, DC: E. De Krafft, 1818), 42, VB363.A2 1818b, Special Collections and Archives, Nimitz Library, United States Naval Academy, Annapolis, MD; U.S. Navy Department, *Annual Report of the Secretary of the Navy, 1830*, 202.

29. Smith, "The Navy before Darwinism," 44; Dick, *Sky and Ocean Joined*, 38–56; Skelton, *An American Profession of Arms*, 98–105; Schlee, *The Edge of an Unfamiliar World*, 24–26; Theberge, *The Coast Survey, 1807–1867*, vol. 1.

30. Hardy, "Matthew Fontaine Maury: Scientist," 402–10.

31. Wilkes, *Autobiography of Rear Admiral Charles Wilkes*, 216–27; *The Army and Navy Chronicle* (Washington, DC: B. Homans, 1837), 4:52. On young naval officers with technical interests see, for example, the naval career of John Dahlgren in Schneller, *A Quest for Glory*, 28–33, 98.

32. Dick, *Sky and Ocean Joined*, 17–44.

33. Nash, *Wilderness and the American Mind*, xi–xiv, 5. On wilderness, see Cronon, "The Trouble with Wilderness, or Getting Back to the Wrong Nature," 7–28; Brady, "The Wilderness of War," 421–47; Brady, *War upon the Land*; Nash, *Wilderness and the American Mind*, 23–43, 67–69; see also Taylor, " 'Wasty Ways': Stories of American Settlement," 291–310. On the frontier archetype in the context of the American West, I have relied on Smith, *Virgin Land*, 61, and Slotkin, *Regeneration through Violence*, 496.

34. Kroll, *America's Ocean Wilderness*, 1–2; Andersson, *Hero of the Atomic Age*, 175.

35. Private Journal of William Reynolds, October 29, 1838, 1:54, transcribed by Thomas Philbrick, Reynolds Family Papers, Archives and Special Collections, Franklin and Marshall College, Lancaster, PA; Maury, *Physical Geography of the Sea*, 262.

36. Henry P. Wells, "The United States Coast and Geodetic Survey," *Harper's Weekly*, October 20, 1888, Papers of George E. Belknap, Miscellaneous Writings, Box 3, Printed Matter 1874–1893, Manuscript Division, Library of Congress (LOC), Washington, DC.; Diary of Robert Weir, *Clara Bell*, Log 164, April 12, 1858, CRC, MSM.

37. Philbrick, *James Fenimore Cooper*, 66; Thoreau, *Cape Cod*, 219–20; Poe, "MS. Found in a Bottle," in *Complete Stories and Poems of Edgar Allan Poe*, 149–50; Melville, *Moby-Dick*, 1214; Philbrick, "Fact and Fiction: The Uses of Maritime History in Cooper's Afloat and Ashore," in Dudley and Crawford, *The Early Republic and the Sea*, 215; Baker, "Mapping and Measuring with Ahab and Wilkes," in *Heartless Immensity*, 43.

## Chapter Two

1. Private Journal of William Reynolds, March 16, 1839, 1:129, transcribed by Thomas Philbrick, Reynolds Family Papers, Archives and Special Collections, Franklin and Marshall College, Lancaster, PA; ibid., March 16, 1839,

2. On the United States Exploring Expedition, see Philbrick, *Sea of Glory*; Stanton, *The Great United States Exploring Expedition*; Viola and Margolis, *Magnificent Voyagers*; Tyler, *The Wilkes Expedition*; Burnett, "Hydrographic Discipline among the Navigators," in Akerman, *The Imperial Map*," 185–259.

3. See Lindgren, "That Every Mariner May Possess the History of the World," 179–205, and Malloy, "Sailors' Souvenirs at the East India Marine Hall," 93–103.

4. Poe, "MS. Found in a Bottle," 155. See also Dupree, *Science in the Federal Government*, 6–9, 44–46. On Antarctic exploration and the American literary imagination, see Wijkmark, "'One of the Most Intensely Exciting Secrets'"; Walden, *The Coastal Frontier and the Oceanic Wilderness*; and Sachs, *Humboldt Current*, 122.

5. Reynolds, *Address on the Subject of a Surveying and Exploring Expedition*, 14; Sachs, *Humboldt Current*, 1–38.

6. Memorial of Edmund Fanning, Senate Document 10, 1833, Zc74 844wix A19, Beinecke Rare Book and Manuscript Library, Yale University, 1–2; Reynolds, *Exploring Expedition. Correspondence between J.N. Reynolds and the Hon. Mahlon Dickerson* (New York, 1838), Zc74 844wix A27, 13, Beinecke Rare Book and Manuscript Library, Yale University, New Haven, CT. See also Sachs, *Humboldt Current*, 118; Reynolds, *Address on the Subject of a Surveying and Exploring Expedition*, 230; and Reynolds, *Voyage of the United States Frigate* Potomac.

7. Reynolds, *Pacific and Indian Oceans*, 284.

8. Kearny, quoted in *The Army and Navy Chronicle* (Washington, DC: B. Homans, 1838), 6:5.

9. Wilkes, *Autobiography of Rear Admiral Charles Wilkes*, 359, 382; R. Semmes to George Foster Emmons, Cincinnati, February 21, 1838, George Foster Emmons Papers, WA MSS 168, Beinecke Rare Book and Manuscript Library, Yale University, New Haven, CT. On John Quincy Adams and American naval science, see Stanton,

*Great United States Exploring Expedition*, 4–5, and Dupree, *Science in the Federal Government*, 39–43.

10. Unidentified newspaper, Norfolk, August 10, 1838, Emmons Papers, WA MSS 166; James K. Paulding to Charles Wilkes, Navy Department, August 11, 1838, in Wilkes, *Narrative*, 1:xxviii–xxix.

11. Paulding to Wilkes, *Narrative*, 1: xxviii–xix.

12. Robinson, *The Coldest Crucible*, 19–20. See also Goetzmann, *Exploration and Empire*; Goetzmann, *New Lands, New Men*; Fabian, *The Skull Collectors*, 121–62; and Adler, "The Capture and Curation of the Cannibal 'Vendovi,'" 255–82.

13. Reynolds, *Exploring Expedition*, 126–27; Wilkes, *Narrative*, 1:429. On John Quincy Adams and American naval science, see Stanton, *Great United States Exploring Expedition*, 4–5; Dupree, *Science in the Federal Government*, 39–43. On the Wilkes Expedition within the broader development of American science and the opportunities missed in forging bonds between civilian and military science, see Schlee, *The Edge of an Unfamiliar World*, 28–36.

14. Bruce, *The Launching of Modern American Science*, 175.

15. Journal of George Foster Emmons, WA MSS 166, Box 2, August 14, 1839, Beinecke Rare Book and Manuscript Library, Yale University; William Reynolds to Father, Mother, et al., *Peacock*, At Sea, September 21, 1840, in Cleaver and Stann, *Voyage to the Southern Ocean*, 182. See also Ehrenberg, Wolter, and Burroughs, "Surveying and Charting the Pacific Basin," in Viola and Margolis, *Magnificent Voyagers*, 164–87.

16. Edney, *Mapping an Empire*, 21–32, 236; Burnett, "Hydrographic Discipline among the Navigators," 203, 201–13.

17. Oliver, *Wreck of the Glide*, 27.

18. Emmons Journal, WA MSS 166, Box 1, August 15, 1839; Journal of Silas Holmes, WA MSS 260, 1:134, Beinecke Rare Book and Manuscript Library, Yale University; Emmons Journal, WA MSS 166, Box 1, August 15, 1839; ibid., August 24, 1839.

19. Jacob, *Sovereign Map*, 187.

20. Wilkes, General Order No. 1, Emmons Journal, WA MSS 166, August 22, 1838; ibid, August 1838; Reynolds Journal, October 25, 1838, 45. See also Holmes Journal, 143, 148, and Rozwadowski, *Fathoming the Ocean*, 177–209.

21. Igler, "On Coral Reefs, Volcanoes, Gods, and Patriotic Geology," 26, 44–49; see also Adler, "The Capture and Curation of the Cannibal 'Vendovi,'" 22. On the expedition's collections, see Philbrick, *Sea of Glory*, 331–45, and Stanton, *Great United States Exploring Expedition*, 364–77.

22. Reynolds Journal, September 16, 1838, 1:53.

23. Palmer, "Antarctic Adventures of the United States Schooner Flying-Fish in 1839," in *Thulia: A Tale of the Antarctic*, 67, 69. On Cook and the Ex. Ex., see Philbrick, *Sea of Glory*, 87–93, and Stanton, *Great United States Exploring Expedition*, 99–104; on metalepsis, see Burnett, *Masters of All They Surveyed,* 39, and

24. Philbrick, *Sea of Glory*, 169–86; Stanton, *Great United States Exploring Expedition*, 159–85; Wilkes, *Narrative*, 2:282.

25. Wilkes to Paulding, U.S.S. *Vincennes*, March 10, 1840, in *The Polynesian*, April 24, 1841, Emmons Journal, WA MSS 166, Box 2.

26. J. N. Reynolds, *Exploring Expedition*, 115; Private Journal of William Reynolds, 2:121, Reynolds Family Papers, Archives and Special Collections, Franklin and Marshall College; Holmes Journal, 289, 300–301. See also Charles Wilkes, *Hydrography*, 23:148–49. On the Salem bêche-de-mer traders, see Dodge, "Fiji Trader," 3–19.

27. Emmons Journal, WA MSS 168, May 20, 1840; Reynolds Private Journal, April 30, 1840, 1:346; William Reynolds to Father, Mother, et al., *Peacock*, At Sea, September 21, 1840, in Cleaver and Stann, *Voyage to the Southern Ocean*, 195.

28. Reynolds, , *Correspondence between J.N. Reynolds and the Hon. Mahlon Dickerson*, 115; Wilkes, *Narrative*, 3:221. On the Salem East India Marine Society and memorials regarding the *Charles Doggett*, see Cleaver and Stann, *Voyage to the Southern Ocean*, 172; Wilkes, *Narrative*, 3:103–5.

29. Lyons, *American Pacificism*, 77; Holmes Journal, 297; Emmons Journal, WA MSS 166, Box 2, May 9, 1840; Reynolds Journal, 2:12; Reynolds Journal, May 16, 1840, 348; Reynolds to Father, Mother, et al., September 21, 1840, in Cleaver and Stann, *Voyage to the Southern Ocean*, 157–58; Reynolds Journal, May 16, 1840, 348. On Fijian cannibalism, I have relied on Obeyesekere, "Cannibal Feasts in Nineteenth-Century Fiji: Seamen's Yarns and the Ethnographic Imagination," in Barker, Hulme, and Iverson, *Cannibalism and the Colonial World*, 63; Obeyesekere, *Cannibal Talk*; and Arens, *The Man-Eating Myth*.

30. Reynolds Journal, August 14, 1840, 1:350.

31. Wilkes, *Narrative*, 3:49–51, 54–59, 408–9.

32. Reynolds to Father, Mother, et al., September 21, 1840, in Cleaver and Stann, *Voyage to the Southern Ocean*, 182. Here, again, I am relying on Burnett, "Hydrographic Discipline among the Navigators."

33. Emmons Journal, WA MSS 166, Box 2, May 19, 1840; Wilkes, *Narrative*, 1:xvii; Journal of Lieutenant George T. Sinclair, May 11, 1840, Records Relating to the U.S. Exploring Expedition, NA 314, Roll 21, National Archives and Records Administration (NARA), Washington, DC; Emmons Journal, WA MSS 166, Box 2, August 11, 1840. See also Burnett, "Hydrographic Discipline among the Navigators," 185–87, 246–59, and Wilkes, *Hydrography*, 8.

34. Wilkes, *Narrative*, 1:xvii. See also Burnett, "Hydrographic Discipline among the Navigators"; Reynolds Journal, August 15, 1840, 2:130; Sinclair Journal, August 15, 1840.

35. Sinclair Journal, May 11, 1840; Emmons Journal, WA MSS 166, Box 1, October 24, 1839. See also Burnett, "Hydrographic Discipline among the Navigators," 248.

36. Reynolds Journal, 2:129; ibid., 1:236; William Reynolds to Father, Mother, et al. September 21, 1840, in Cleaver and Stann, *Voyage to the Southern Ocean*, 183. See also Stanton, *Great United States Exploring Expedition*, 196–97.

37. Adler, "The Capture and Curation of the Cannibal 'Vendovi,'" 22. On Ro Veidovi, see Fabian, *The Skull Collectors*, 121–62.

38. Emmons Journal, WA MSS 166, Box 2, June 9, 1840; Short, *Cartographic Encounters*, 9–19. On indigenous agency and cross-cultural encounters, I am relying on Dening, *Islands and Beaches*; Dening, "Deep Time, Deep Spaces: Civilizing the Sea," in Bernhard Klein and Gesa Mackenthun, *Sea Changes*, 13–36; on the maritime prowess of Fijians, see Nuttall, D'Arcy, and Philp, "Waqa Tabu—Sacred Ships: The Fijian *Drua*," 229, 446–49.

39. Obeyesekere, "Cannibal Feasts in Nineteenth-Century Fiji," 63; Obeyesekere, *Cannibal Talk*, 192; Arens, *The Man-Eating Myth*, 9; for a contrary argument, see Sahlins, "Artificially Maintained Controversies," 3–5; Reynolds Journal, 2:12; Wilkes, *Narrative*, 3:234–35; Journal of Frederick Stuart, May 8, 1840, "Remarks on Bitileb," Records Relating to the U.S. Exploring Expedition, NA 313, Roll 20, NARA; Ann Fabian, *The Skull Collectors*, 129–30.

40. Journal of George M. Colvocoresses, WA MSS 101, Box 1, July 12, 1840, 162, Beinecke Rare Book and Manuscript Library, Yale University; James K. Paulding to Charles Wilkes, Navy Department, August 11, 1838, in Wilkes, *Narrative*, 1:xxviii. On constructions of ocean space in Micronesia, which, it should be noted is not Fijian Melanesia, see Steinberg, *The Social Construction of the Ocean*, 39–67.

41. Wilkes, Narrative, 3: 424; Lisa M. Brady, "The Wilderness of War," 421–47; Brady, *War upon the Land*.

42. On casualties, see Wilkes, *Narrative*, 3:281, and Emmons Journal, WA MSS 166, Box 2, July 28, 1840; Holmes Journal, WA MSS 260, vol. 2, July 26, 1840; Wilkes, *Narrative*, 3:281, 284–85.

43. "The United States Exploring Expedition," *Hunt's Merchants' Magazine* 12 (May 1845): 445; Philbrick, *James Fenimore Cooper*, 203–59, 220–21. See also, Gates, "Cooper's Sea Lions and Wilkes' Narrative," 1069–1075; Gates, "Cooper's Crater and Two Explorers," 243–46; Philbrick, *Sea of Glory*, 339.

44. Merk, *The Oregon Question*, 211–12; Bancroft to Palmerston, Eaton Square, July 31, 1848, in House of Representatives, *Executive Documents*, 42nd Cong., 3d Sess. (1872), at 250; Merk, *The Oregon Question*, 211.

45. Jacob, *The Sovereign Map*, 205; Official Report of Lieutenant Wilkes, *The Polynesian*, April 24, 1841, Emmons Journal, WA MSS 166, Box 2; Emmons Journal, May 17, 1840, WA MSS 166, Box 2; Philbrick, *Sea of Glory*, 362.

46. Davis, "The United States Exploring Expedition," 83; Reynolds Journal, July 26, 1840, 2:19. On space and meaning, I have relied on Tuan, *Space and Place*.

47. *The Liberator*, June 15, 1849, in Ward, *American Activities in the Central Pacific, 1790–1870*, 2:448; Williams and Calvert, *Fiji and the Fijians*, 10.

48. Twain, *The Autobiography of Mark Twain*, 1:367. On Wilkes's court martial, see Philbrick, *Sea of Glory*, 307–30, and Burnett, "Hydrographic Discipline among the Navigators," 246–59. The riverboat pilot's call of "Mark, Twain!" indicated that the river bottom was two fathoms from the surface, shallow enough for light draft riverboats to run aground.

## Chapter Three

1. Whitman, *Leaves of Grass*, 56; *Daily Alta California*, August 16, 1850; ibid., August 12, 1853; *New York Daily Times*, February 10, 1854. See also "Steamships at the Port of San Francisco," The Maritime Heritage Project, 1846–1899, http://www.maritimeheritage.org/ships/ssSanFrancisco.htm (accessed January 5, 2017).

2. Maury, *Explanations and Sailing Directions*, 8th ed., 1:108; Maury, *The Physical Geography of the Sea*, 57–58.

3. On Maury, I have relied on Williams, *Matthew Fontaine Maury*; Towle, "Science, Commerce and the Navy on the Seafaring Frontier"; Lewis, *Matthew Fontaine Maury*; Grady, *Matthew Fontaine Maury*; Reingold, "Two Views of Maury . . . and a Third," 370–72; see also a special forum on Maury in the *International Journal of Maritime History* 28 (May 2016): 388–420.

4. Maury, "On the Navigation of Cape Horn," 54–55.

5. Maury, *A New Theoretical and Practical Treatise on Navigation*, v.

6. Maury to Richard L. Maury, Philadelphia, October 29, 1835, Box 1, Papers of Matthew Fontaine Maury, MSS31682, LOC, Washington, DC.

7. "Opinions of Navigators and Professors," in Maury, *Treatise on Navigation*, 1–2, 5.

8. Henry to Titus W. Powers, March 13, 1836, New York, in Reingold, *The Papers of Joseph Henry*, 3:23. See also Dick, *Sky and Ocean Joined*, 111–12.

9. Maury to Miss Maury, Fredericksburg, Virginia, April 5, 1838, Box 65, Folder 2, Maury Papers, LOC; Burnett, "Matthew Fontaine Maury's 'Sea of Fire,'" 121. See also Williams, *Matthew Fontaine Maury*, 197–202, 116–19.

10. Maury, "An Appeal to the Agricultural Interests of Virginia," Box 5, Maury Papers, LOC; Maury to Robert Walsh, Washington, DC, January 24, 1848, Records of the Naval Observatory, Record Group (RG) 78, Letters Sent, vol. 2, NARA.

11. William M. Crane, Chief of the Bureau of Ordnance and Hydrography, Circular, December 16, 1842, Records of the Hydrographic Office, RG 37, Miscellaneous Letters Sent, NARA; Maury to Robert Walsh, Washington, DC, December 30, 1847, RG 78, Letters Sent, vol. 2, NARA.

12. Maury to Dr. Drake, Washington, DC, September 30, 1845, RG 78, Letters Sent, vol. 2, NARA; M. F. Maury to Dr. [Daniel] Drake, Hydrographical Office, September 30, 1845, RG 78, Letters Sent, vol. 2, NARA; Williams, *Matthew Fontaine Maury*, 153, 61n521, 32n532.

13. Maury, *Explanations and Sailing Directions*, 3rd ed., 23; Maury to Julius Rockwell, Washington, DC, January 15, 1849, RG 78, Letters Sent, vol. 3, NARA; Maury, "Blank Charts on Board Public Cruisers," 459; Maury to Matthew Maury [cousin], Washington, DC, July 14, 1848, Maury Papers, General Correspondence, Box 3, LOC; Maury to the Owners and Masters of the New Bedford and New London Whaleships, Washington, DC, January 15, 1849, RG 78, Letters Sent, vol. 3, NARA.

14. Maury to H[enry] Mactier Warfield, Washington, D.C., June 1, 1848, Records of the Naval Observatory, RG78, Letters Sent, Vol. 3, NARA.

15. Williams, *Matthew Fontaine Maury*, 180; Maury, *Explanations and Sailing Directions*, 5:250–51, 24, 270–71.

16. Maury, *Physical Geography of the Sea*, 25. On the Humboldtian connections between American science and literature in the nineteenth century, see Walls, *The Passage to Cosmo*.

17. Maury, *Physical Geography of the Sea*, ix–xi. Maury, "Address Delivered before the Geological and Mineralogical Society of Fredericksburg, May 18, 1836," Box 1, Maury Papers, LOC; Maury, *Physical Geography of the Sea*, ix–xi.

18. Maury to R. Kennedy, Washington, DC, November 11, 1844, RG 78, Letters Sent, vol. 1, NARA. The American scientist William Redfield had tracked and studied the paths of hurricanes during the 1820s by examining ship logs.

19. Abstract Log of Brig *Georgiana*, Jonathan Chase, Mozambique to New York, 1851, M1160, vol. 27, Roll 9, RG27, NARA; Abstract Log of Ship *Contest*, William Brewster, New York to San Francisco, 1852, M1160, vol. 114, Roll 31, RG27, NARA; Maury, *Physical Geography of the Sea*, vii.

20. Maury to Robert Walsh, Washington, DC, July 9, 1847, RG 78, Letters Sent, vol. 2, NARA; Maury, "An Appeal to the Agricultural Interests of Virginia," Maury Papers, General Correspondence, Box 5, LOC; Captain Leslie Bryson, quoted in Maury, *Explanations and Sailing Directions*, 281; Captain Smyley, quoted in Maury, *Explanations and Sailing Directions*, 287; Burnett, "Hydrographic Discipline among the Navigators," in Akerman, *The Imperial Map*, 248.

21. Maury to R[obert] B. Forbes, Washington, DC, July 16, 1848, RG 78, Letters Sent, vol. 2, NARA; Maury to John Y. Mason, Washington, DC, September 1848, RG 78, Letters Sent, vol. 3, NARA.

22. Maury to Ann Maury, Washington, DC, April 19, 1848, Maury Papers, General Correspondence, Box 3, LOC; Maury, *Physical Geography of the Sea*, x; Maury to John Y. Mason, Washington, DC, September 1848, RG 78, Letters Sent, vol. 3, NARA.

23. Maury to Captain Thomas Freeman, Washington, DC, March 29, 1851, RG 78, Letters Sent, vol. 5, NARA; Maury, *Explanations and Sailing Directions*, 18; Maury to John Y. Mason, Washington, DC, September 1848, RG 78, Letters Sent, vol. 3, NARA; the Bible passage referenced is Ps. 107.

24. Maury to Captain Frank Smith, Washington, DC, January 1, 1849, RG 78, Letters Sent, vol. 3, NARA.

25. Maury to William Blackford, Washington, DC, March 12, 1849, Box 3, Maury Papers, LOC. On the commodification of whale bodies, see Nathaniel Philbrick, *In the Heart of the Sea*, 65.

26. Heflin, *Herman Melville's Whaling Years*; Melville, *Moby-Dick*, 1003–4. On Melville's critique of science, see Baker, "Mapping and Measuring with Ahab and Wilkes," in *Heartless Immensity*, 30–44.

27. Maury, *Explanations and Sailing Directions*, 57, 133.

28. Ibid., 8th ed., 1:308. On Beaufort, see Ritchie, *Admiralty Chart*, 189–99.

29. Maury to the Owners and Masters of the New Bedford and New London Whaleships, National Observatory, January 15, 1849, RG 78, Letters Sent, vol. 3, NARA.

30. Williams, *Matthew Fontaine Maury*, 190–91; Maury, "Paper on the Currents of the Sea as Connected with Geology, Read before the Association of American Geologists and Naturalists, May 14, 1844," quoted in *Explanations and Sailing Directions*, 121.

31. Maury, *Explanations and Sailing Directions*, 19, 24; Maury to Robert Walsh, Washington, DC, January 24, 1848, RG 78, Letters Sent, vol. 2, NARA; Maury, *Explanations and Sailing Directions*, 19; Maury to George Manning, Washington, DC, November 13, 1848, RG 78, Letters Sent, vol. 3, NARA; Maury, *Explanations and Sailing Directions*, 24.

32. Abstract Log of the Ship *Venice*, quoted in Maury, *Explanations and Sailing Directions*, 294; Abstract Log of the Ship *Creole*, Captain Foster, Le Havre to New Orleans, 1849, MSS 1160, vol. 80, Roll 23, RG27, NARA; Phinney, quoted in Maury, *Physical Geography of the Sea*, 8.

33. Maury to R[obert] B. Forbes, April 27, 1848, RG 78, Letters Sent, vol. 3, NARA; Maury to John Q. Adams, Washington, DC, November 17, 1847, RG 78, Letters Sent, vol. 2, NARA; Maury to E. and G. W. Blunt, Washington, DC, October 28, 1846, RG 78, Letters Sent, vol. 2, NARA; Maury to Sherman and Smith, Washington, DC, February 3, 1848, RG 78, Letters Sent, vol. 2, NARA.

34. Maury, *Explanations and Sailing Directions*, 232, 271, 240.

35. See Burstyn, "Seafaring and the Emergence of American Science," 101; Dick, *Sky and Ocean Joined*, 107–9, 111–12; Williams, *Matthew Fontaine Maury*, 168–175, 203–5, 235–36; and Schlee, *The Edge of an Unfamiliar World*, 39.

36. Maury, *Explanations and Sailing Directions*, 42, 47–48; Maury, *Physical Geography of the Sea*, 63. I am drawing heavily on Burnett, "Matthew Fontaine Maury's 'Sea of Fire,'" in Driver and Martins, *Tropical Visions in the Age of Empire*, 131, 47n243–44; Marx, *Machine in the Garden*, 160–64, 190–96.

37. Maury, *Explanations and Sailing Directions*, 57; Maury to the Editor of *The New York Evangelist*, Washington, DC, January 22, 1855, Maury Papers, General Correspondence, Box 4, LOC; the Bible verses that Maury cited are Job 26:7 and Eccles. 1:7, among others; on Dana, see Igler, "On Coral Reefs, Volcanoes, Gods, and Patriotic Geology," 44–45; Maury, *Explanations and Sailing Directions*, 45; Dupree, "Christianity and the Scientific Community in the Age of Darwin," in Lindberg and Numbers, *God and Nature*, 351–68; Burnett, "Matthew Fontaine Maury's 'Sea of Fire,'" 129–30.

38. Burnett, "Matthew Fontaine Maury's 'Sea of Fire,'" 127; Letter to the Editor of *The Boston Atlas*, February 18, 1857, Maury Papers, General Correspondence, Box 7, LOC. See also James Rodger Fleming, Meteorology in America, 1800–1870 (Baltimore: Johns Hopkins University Press, 1990), 81–93; Williams, *Matthew Fontaine Maury*, 204.

39. Leighly, introduction to *The Physical Geography of the Sea and Its Meteorology*, by Matthew Fontaine Maury, xx, xvii–xix.

40. Excerpts from the *Southern Literary Messenger*, July 1841, Box 2, Maury Papers, LOC; Maury to M. Lomard, Washington, DC, January 2, 1856, Box 5, Maury Papers, LOC.

41. Matthew Fontaine Maury to Ann Maury, Observatory, Washington, DC, June 30, 1850, Maury Papers, General Correspondence, Box 4; On the Lynch Expedition to the Dead Sea, see Jampoler, *Sailors in the Holy Land*; on the North Pacific Exploring Expedition, see Rozwadowski, *Fathoming the Ocean*, 50–57; Karp, "Slavery and Sea Power," 293, 319; Hecht, *The Scramble for the Amazon*, 142–54; and Ponko, *Ships, Seas, and Scientists*, 206–30.

42. Maury, "Programme of Instructions for the 'Taney,'" drawn by request, for the Secretary of the Navy, October 4, 1849, RG 78, Letters Sent, vol. 4, NARA. See also Maury, *Explanations and Sailing Directions*, 62.

43. Maury to Franklin Minor, Washington, DC, October 16, 1856, Maury Papers, General Correspondence, Box 6, LOC; Maury, *Explanations and Sailing Directions*, 90. See also George M. Brooke, *John Mercer Brooke: Naval Scientist and Educator*, 57–59; Rozwadowski, *Fathoming the Ocean*, 84–86; and Schlee, *Edge of an Unfamiliar World*, 53–54.

44. Williams, *Matthew Fontaine Maury*, 231–32; Maury, quoted in Rozwadowski, *Fathoming the Ocean*, 90, 92; *New York Herald* and "The Recent Soundings for the Atlantic Telegraph," *Illustrated London News* (Fall 1857), quoted in Rozwadowski, *Fathoming the Ocean*, 90.

45. George M. Brooke, *John Mercer Brooke*, 58, 84; Lieutenant William Rogers Taylor, quoted in Maury, *Explanations and Sailing Directions*, 80; Maury, "Blank Charts on Board Public Cruisers," 459; Maury, "Suggestions for the Attention of the Home Squadron," October 3, 1843, RG 78, Letters Sent, vol. 1, NARA.

46. Maury to Bishop Joseph Harvey Utley, University of Virginia, September 20, 1855, Maury Papers, General Correspondence, Box 5, LOC. See also Williams, *Matthew Fontaine Maury*, 269–308.

47. Maury, *Physical Geography of the Sea*, x. See also Rozwadowski, *Fathoming the Ocean*, 44.

## Chapter Four

1. Cummings, *A Synopsis of the Cruise of the U.S.S. "Tuscarora,"* 20.

2. "An Important Discovery by the Pacific Exploring Expedition—A 'Telegraph Plate' for the Pacific Cable," Belknap Papers, MSS52696, Box 3, Miscellaneous Writings, Clippings, 1891–99 and undated, LOC, Washington, DC. See also LaFeber, *The New Empire*, and Hagan, *American Gunboat*.

3. Cummings, *Synopsis of the Cruise*, 60.

4. On the U.S. Navy in the Civil War, see McPherson, *War on the Waters*, and Still, Taylor, and Delaney, *Raiders and Blockaders*; on the Mexican-American War, see Bauer, *Surfboats and Horse Marines*, and Morison, *"Old Bruin."*

5. Semmes, *Memoirs of Service Afloat*, 579–81.

6. On the fate of the American merchant and whaling fleets during and after the Civil War, see Roland, Bolster, and Keyssar, *The Way of the Ship*, 196, 198.

7. Seward, quoted in Paolino, *Foundations of the American Empire*, 207; Cong. Globe, 32nd Cong., 1st Sess. (July 29, 1852, 1973); Cong. Globe, 40th Cong., 1st Sess. (July 1, 1868), at 3660–61; ibid., 3659–60. See also Greenberg, *Manifest Manhood*, 69–70; Love, *History of the U.S. Navy*, 1:324; and Cumings, *Dominion from Sea to Sea*; for antebellum Pacific empire, see Igler, *The Great Ocean*.

8. Hagan, *American Gunboat Diplomacy*, 8; Porter, quoted in Hagan, *American Gunboat Diplomacy*, 23; U.S. Navy Department, *Annual Report of the Secretary of the Navy, 1868*, 21–22; U.S. Navy Department, *Annual Report of the Secretary of the Navy, 1869*, 63. See also Shufeldt, *The Relation of the Navy to the Commerce of the United States*; U.S. Navy Department, *Annual Report of the Secretary of the Navy, 1878*, 11–16; Drake, *Biography of Rear Admiral Robert Wilson Shufeldt*; Drake, "Robert Wilson Shufeldt," in Bradford, *Captains of the Old Steam Navy*, , 275–300. On naval officers, business, and expansion in this era, see Karsten, *The Naval Aristocracy*, 187–249; on masculinity and American empire, see Greenberg, *Manifest Manhood*, and Kristin L. Hoganson, *Fighting for American Manhood*.

9. U.S. Navy Department, *Annual Report of the Secretary of the Navy, 1868*; U.S. Navy Department, *Annual Report of the Secretary of the Navy, 1870*, 7.

10. See U.S. Navy Department, *Annual Report of the Secretary of the Navy, 1885*, 24, and Pinsel, *150 Years of Service on the Sea*, 1:16.

11. On USS *Saginaw*, I have relied on Van Tilburg, *A Civil War Gunboat in Pacific Waters*, 9–36.

12. R. W. Meade, "Alaska," *Appleton's Journal*, July 22, 1871, reprinted in Robert N. DeArmond, *The USS Saginaw in Alaska Waters, 1867–1868*, 132; U.S. Navy Department, *Annual Report of the Secretary of the Navy, 1868*, 16.

13. Van Tilburg, *Civil War Gunboat in Pacific Waters*, 158; Diary of Peveril Meigs, quoted in DeArmond, *Saginaw in Alaska Waters*, 53 (hereafter referred to as Meigs Diary); Van Tilburg, *A Civil War Gunboat in Pacific Waters*, 153; Baker, *Geographic Dictionary of Alaska*, 40.

14. Meigs Diary, 83; Meade, "Alaska," *Appleton's Journal*, July 22, 1871, reprinted in Robert N. DeArmond, *The USS Saginaw in Alaska Waters, 1867–1868*, 131; Meade, "Alaska," *Appleton's Journal*, July 29, 1871; Meade, "Alaska," *Appleton's Journal*, July 22, 1871, 124, 129–30. On Russia, empire, and the natural world, I have relied on Jones, *Empire of Extinction*.

15. Meade to T. T. Craven, January 2, 1869, in DeArmond, *Saginaw in Alaska Waters*, 61–62; Meade, "Alaska," *Appleton's Journal*, July 29, 1871, 134; Meigs Diary, 60; Schroeder, *A Half Century of Naval Service*, 19.

16. Meade to T .T. Craven, February 24, 1869, in DeArmond, *Saginaw in Alaska Waters*, 92; Schroeder, *Half Century of Naval Service*, 19; See Boot, *The Savage Wars of Peace*, 30–39.

17. Charles Walcott Brooks, "Our Furthest Outpost," *Old and New*, June 1870, 828–38; Taylor, "The American Navy in Hawai'i," 907–33. On Midway, I have relied on Van Tilburg, *Civil War Gunboat*, 188–217.

18. Van Tilburg, *Civil War Gunboat*, 198–204. See also Montgomery Sicard to Rear Admiral Thomas Turner, "Deepening Channel at Midway Islands," No. 16, in U.S. Navy Department, *Annual Report of the Secretary of the Navy, 1871*, 225–34.

19. Brooks, "Our Furthest Outpost," *Old and New* 1 (June 1870): 829, 835.

20. U.S. Navy Department, *Annual Report of the Secretary of the Navy, 1871*, 225, 228, 230, 231.

21. Ibid., 229; U.S. Navy Department, *Annual Report of the Secretary of the Navy, 1870*, 8–9. Read, *The Last Cruise of the Saginaw*, 10.

22. Read, The Last Cruise of the Saginaw, 5, 12–13.

23. Ibid., 24, 20–21. See also Van Tilburg, *A Civil War Gunboat*, 222–79.

24. U.S. Navy Department, *Annual Report of the Secretary of the Navy, 1869*, 24; Thomas O. Selfridge, "Report of the Survey of the Isthmus of Darien," No. 14, in U.S. Navy Department, *Annual Report of the Secretary of the Navy, 1871*, 178. On the canal debate in the navy, see Stockton, "The Commercial Geography of the American Inter-Oceanic Canal," 75–93; Ammen, "Proceedings in the General Session of the Canal Congress," 153–85; Hagan, *American Gunboat Diplomacy*, 143–59. Historians espousing the "Dark Ages" interpretation of this period in U.S. naval history, largely reflecting Alfred Thayer Mahan's influence, are, among others, Sprout, *The Rise of American Naval Power*, Sloan, *Benjamin Franklin Isherwood, Naval Engineer*; for revisions to this interpretation, see Hagan, *American Gunboat Diplomacy*; Buhl, "Maintaining an 'American Navy,' 1865–1889," in Hagan, *In Peace and War*, 145–85; Buhl, "Mariners and Machines," 703–72; and Rentfrow, *Home Squadron*. Despite more recent scholarship to the contrary, there remains a general belief that this era of naval history was one of stagnation.

25. Thomas O. Selfridge, "Report of the Survey of the Isthmus of Darien for 1873," No. 12, in U.S. Navy Department, *Annual Report of the Secretary of the Navy, 1873*, 173.

26. E. P. Lull to Robeson, Washington, DC, October 25, 1873, in U.S. Navy Department, *Annual Report of the Secretary of the Navy, 1873*, 202–4.

27. Robinson, *The Coldest Crucible*, 6, 10.

28. See *How to Dispose of the Remains of Our Navy* (lithograph by Mayer, Merkel, and Ottmann), *Puck* 11 (May 17, 1882); "Report to the President of the United States of the Action of the Navy Department in the Matter of the Disaster to the United States Exploring Expedition toward the North Pole, Accompanied by a Report of the Examination of the Rescued Part, Etc.," No. 18, in U.S. Navy Department, *Annual Report of the Secretary of the Navy, 1873*, 284. On the origins of the *Puck* cartoon in Landseer's painting, I wish to thank Kelly P. Bushnell.

29. Geo. F. Robeson, Spencer F. Baird, Wm. Reynolds, H. W. Howgate to the president, Washington, DC, June 16, 1873, in U.S. Navy Department, *Annual Report of the Secretary of the Navy, 1873*, 287; Examination of Dr. Emil Bessels, Washington, DC, October 16, 1873, "Examination Conducted at Washington, DC, of the Party Separated on the Ice from the United States Steamer Polaris Expedition toward the North Pole," in U.S. Navy Department, *Annual Report of the Secretary of the Navy, 1873*, 546;

U.S. Navy Department, *Annual Report of the Secretary of the Navy, 1873*, 283; Robeson et al. to the president, June 16, 1873, in U.S. Navy Department, *Annual Report of the Secretary of the Navy, 1873*, 294.

30. "Examination Conducted at Washington, D.C., of the Party Separated on the Ice from the United States Steamer Polaris Expedition toward the North Pole," in U.S. Navy Department, *Annual Report of the Secretary of the Navy, 1873*, 546; Joseph Henry to Robeson, Washington, DC, June 9, 1871, in U.S. Navy Department, *Annual Report of the Secretary of the Navy, 1871*, 241. On the Polaris Expedition, see Robinson, *Coldest Crucible*, 55–82; see also Davis, *Narrative of the North Polar Expedition*.

31. U.S. Navy Department, *Annual Report of the Secretary of the Navy, 1883*, 126; Sachs, *The Humboldt Current*, 286, 292. On the Jeanette Expedition, see Guttridge, *Icebound: The* Jeanette *Expedition's Quest for the North Pole*, and Sides, *In the Kingdom of Ice*; see also Melville, *In the Lena Delta*.

32. Schley, *Forty-Five Years under the Flag*, 432. On technology in the late nineteenth century, I have relied on Pursell, *The Machine in America*; Kasson, *Civilizing the Machine*; on technological change in the navy in this era, see McBride, *Technological Change and the United States Navy*; Herrick, *The American Naval Revolution*.

33. Endicott, *Wrecked among Cannibals in the Fijis*, 18; Rozwadowski, *Fathoming the Ocean*, 27; Twain, *Following the Equator*, 94; Twain, *The Autobiography of Mark Twain*, 1:367; and Rozwadowski, *Fathoming the Ocean*, 62. See Verne, *Twenty Thousand Leagues under the Sea*.

34. Unidentified Newspaper, Papers of George E. Belknap, MSS52696, Box 3, Clippings, 1891–1899 and undated, LOC; George E. Belknap, "Something about Deep-Sea Sounding, I," in *The United Service*, 1:161; Belknap, "Something about Deep-Sea Sounding, II," 349.

35. Rear Admiral Geo. E. Belknap, Brookline, Massachusetts, November 18, 1895, "Deep Sea Sounding: Admiral Belknap on the Work of the Penguin and the Tuscarora," Unidentified and Undated Newspaper, Belknap Papers, MSS52696, Box 3, Clippings, 1891–1899 and undated, LOC; Belknap, "Something about Deep-Sea Sounding, " 359; Kasson, *Civilizing the Machine*, 139–42. On the Thomson Deep Sea Sounder, see Belknap, "Something about Deep-Sea Sounding, I," 179–80; on the challenges of sounding and sounder design, see Rozwadowski, *Fathoming the Ocean*, 69–95.

36. Rozwadowski and van Keuren, *The Machine in Neptune's Garden*, xiv–xv; Marx, *The Machine in the Garden*; Judd, *Second Nature*; Belknap, "Something about Deep-Sea Sounding, II," 356; Belknap, "Something about Deep-Sea Soundings, I," 178; Jewell, "Deep Sea Sounding," 62.

37. Rozwadowski, *Fathoming the Ocean*, 168–209; Belknap, "Something about Deep-Sea Sounding," 364–65, 356; On *Tuscarora*'s voyage, see Belknap, "Something about Deep-Sea Sounding, II," 362, 371; on the Challenger Expedition, see Schlee, *Edge of an Unfamiliar World*, 143–62.

38. Rozwadowski, *Fathoming the Ocean*, 62; "Secrets of the Deep" and "Deep Sea Searches Rendered Easy," Unidentified and Undated Newspapers, Belknap Papers,

MSS52696, Box 3, Clippings, 1891–1899 and undated, LOC; "How Deep Sea Soundings Are Made," *New York Herald*, undated, Belknap Papers, MSS52696, Box 3, Clippings, 1891–1899 and undated, LOC; "The Pacific Deep Sea Soundings," Unidentified and Undated Newspaper, Belknap Papers, MSS52696, Box 3, Clippings, 1891–1899 and undated, LOC; Unidentified and Undated Newspaper, Yokohoma, Japan, May 15, 1874, Papers of George E. Belknap, MSS52696, Box 3, Miscellaneous Writings, Clippings, 1891–99 and undated, LOC; Cummings, *Cruise of the "Tuscarora,"* 60.

39. Wyville Thomson, quoted in Belknap, "Something about Deep-Sea Soundings," 176; Sigsbee to Belknap, U.S. Hydrographic Office, Washington, DC, February 10, 1879, Belknap Papers, MSS52696, Box 1, Correspondence, 1879, LOC. On the scientific relationship between the Coast Survey and the navy, see Alaniz, "Dredging Evolutionary Theory"; see also Manning, *U.S. Coast Survey vs. Naval Hydrographic Office.*

40. Sigsbee, *Deep-Sea Sounding and Dredging,* 12.

41. Patterson, "Note by the Superintendent," in Sigsbee, *Deep-Sea Sounding and Dredging,* 4; Agassiz, *Three Cruises of the United States Coast and Geodetic Survey Steamer "Blake,"* xxii; Sigsbee to Belknap, U.S. Hydrographic Office, Washington, DC, February 10, 1879, Belknap Papers, MSS52696, Box 1, Correspondence, 1879, LOC; Sigsbee, *Deep-Sea Sounding and Dredging,* 25. On the prevailing argument that this was an era of stagnation in naval science with little to offer the field of oceanography, see Schlee, *Edge of an Unfamiliar World,* 63, 79.

42. On *Tuscarora* and the *Virginius* Affair, see Cummings, *Cruise of the "Tuscarora,"* 24; on *Tuscarora* in Hawaii, see Cummings, *Cruise of the "Tuscarora,"* 36–39; U.S. Navy Department, *Annual Report of the Secretary of the Navy, 1874,* 8.

43. "The Kanaka Isles," Unidentified and Undated Newspaper, Belknap Papers, MSS52696, Box 3, Clippings, 1891–1899 and undated, LOC; Headrick, *The Tentacles of Progress,* 98; Headrick, *The Tools of Empire,* 163; "Debate on Annexation," Unidentified and Undated Newspaper, Belknap Papers, MSS52696, Box 3, Clippings, 1891–1899 and undated, LOC; Belknap, unpublished manuscript, Belknap Papers, MSS52696, Box 3, Miscellaneous Writings, LOC; "Anent Jingoes," Unidentified and Undated Newspaper, Belknap Papers, MSS52696, Box 3, Clippings, 1891–1899 and undated, LOC; see also Hunt, "Doing Science in a Global Empire: Cable Telegraphy and Electrical Physics in Victorian Britain," in Lightman, *Victorian Science in Context,* 312–33; Kennedy, "Imperial Cable Communications and Strategy," 728–52; Reidy, Kroll, and Conway, *Exploration and Science,* 156.

44. Wilkes, *Narrative,* 4:79.

45. On the Treaty of Reciprocity, see Tansill, *Diplomatic Relations between the United States and Hawaii*; J. G. Walker to the Secretary of the Navy, U.S. Flagship *Philadelphia*, Honolulu, Hawaiian Islands, June 21, 1894, in Letter from the Acting Secretary of the Navy, S. Doc. No. 42, 53rd Cong. 3rd Sess., at 5.

## Chapter Five

1. Proceedings of a Court of Inquiry Convened to Investigate the Grounding of the U.S.S. *Albany*, Case 4941, RG 125, Records of the Office of the Judge Advocate General (Navy), Box 73, NARA, Washington, DC.

2. See Mahan, *The Influence of Sea Power*, 28; Mahan, "A Twentieth-Century Outlook," 531–32; Mahan, "Hawaii and Our Future Sea-Power," 4; Mahan, "The Isthmus and Sea Power," 459–72; Mahan, "Hawaii and Our Future Sea-Power," 8; Mahan, *The Influence of Sea Power*, 138; and Tucker and Russell, *Natural Enemy, Natural Ally*, 1–14.

3. See Hagan, *This People's Navy*; on the so-called Naval Revolution in the United States, see Shulman, *Navalism and the Emergence of American Sea Power*; Karsten, *The Naval Aristocracy*; Herrick, *The American Naval Revolution*; and Rentfrow, *The Home Squadron*.

4. Seager, *Alfred Thayer Mahan*, 218; Mahan, *The Influence of Sea Power upon History*, 138. On Mahan, see Seager, *Alfred Thayer Mahan*; Seager, "Ten Years before Mahan," 491–512; Hattendorf, *The Influence of History on Mahan*; Hattendorf, *Mahan on Naval Strategy*; Livezey, *Mahan on Sea Power*; Sumida, *Inventing Grand Strategy and Teaching Command*.

5. Mahan, "Discussion of the Elements of Sea Power," in *The Influence of Sea Power upon History*, 25–89; Mahan to Roy B. Marston, New York, February 19, 1897, in Seager and Maguire, *Letters and Papers of Alfred Thayer Mahan*, 2:493. On misinterpretations of Mahan, see Sumida, *Inventing Grand Strategy and Teaching Command*, 1–5, 99–117.

6. Mahan, *The Influence of Sea Power upon History*, 28-9, 35; Mahan, "A Twentieth-Century Outlook," 531–32; Mahan, "Hawaii and Our Future Sea-Power," 4; Mahan, "The Isthmus and Sea Power," 459–72; Mahan, "Hawaii and Our Future Sea-Power," 8; Mahan, *The Influence of Sea Power upon History*, 138.

7. Seager, *Alfred Thayer Mahan*, 193–97, 217–19, 247, 261–62, 190.

8. Chadwick, *The Relations of the United States and Spain*, 1:17.

9. Proceedings of a Court of Inquiry Convened on Board the United States Ship *Iowa*, Case 4904, RG 125, Records of the Office of the Judge Advocate General (Navy), Box 65, NARA; Francis J. Higginson to Theodore Roosevelt, U.S.S. *Massachusetts*, Dry Tortugas, February 11, 1898, in Allen, *The Papers of John Davis Long, 1897–1904*, 46–47. On the professional stakes for officers running their ships aground, see Article 4, Subsection 10, *Regulations for the Government of the United States Navy, 1896*, 432, RG 80, General Records of the Department of the Navy, Entry 35, PC 31, NARA.

10. Chadwick, *Relation of the United States with Spain*, 8; Higginson to Roosevelt, February 11, 1898, *The Papers of John Davis Long*, 46–47; Trask, *War with Spain*, 88.

11. Mahan, "Blockade in Relation to Naval Strategy," 856.

12. See Trask, *War with Spain*, 90–91, 108–11, 120, and Wiley, *An Admiral from Texas*, 81.

13. Austin M. Knight to Bradford, U.S.S. *Yankton*, Santo Cruz del Sur, Cuba, February 6, 1903, RG 80, General Records of the Department of the Navy, General Board, Box 28, Folder 404-2, NARA. On Bradford's suspicion of Spanish hydrographic intentions, see R. B. Bradford, 2nd Endorsement, January 20, 1903, RG 19, Bureau of Equipment, Correspondence, 1899–1910, Box 101, NARA.

14. J. E. Craig to George Grantham Bain, Washington, DC, August 13, 1898, RG 37, Records of the Hydrographic Office, Entry 32, Letters Sent and Received, February 1885–December 1901, Box 111, NARA; Spencer S. Wood to John D. Long, U.S.S. *Du Pont*, At Sea, May 7, 1898, RG 37, Records of the Hydrographic Office, Entry 32, Letters Sent and Received, February 1885–December 1901, Box 108, NARA; George C. Remey to Royal B. Bradford, Key West, June 29, 1898, RG 37, Records of the Hydrographic Office, Entry 32, Letters Sent and Received, February 1885–December 1901, Box 110, NARA.

15. Craig to Royal B. Bradford, Washington, DC, September 29, 1898, in U.S. Navy Department, *Annual Report of the Secretary of the Navy for the Year 1898*, 289; J. C. Thomas to John D. Long, Corsicana, Texas, May 1898, RG 37, Records of the Hydrographic Office, Entry 32, Letters Sent and Received, February 1885–December 1901, Box 108, NARA. On the expanding role of geography and maps in American culture and as they specifically relate to American empire, see Schulten, *The Geographical Imagination in America*, and Smith, *American Empire*.

16. Wiley, *Admiral from Texas*, 81–86.

17. Ibid., 87, 81.

18. Bernadou, "The 'Winslow' at Cardenas," 701. A copy of *Sailing Directions, Caribbean Sea and Gulf of Mexico*, vol. 1, no. 86, 1898, is reprinted in U.S. War Department, *Military Notes on Cuba*, 310.

19. Bernadou, "'Winslow' at Cardenas," 301; John B. Bernadou to Long, Key West, May 16, 1898, in U.S. Navy Department, *Appendix Bureau of Navigation 1898*, 202; Bernadou, "'Winslow' at Cardenas," 301; Bernadou to Long, Key West, May 16, 1898, in *Appendix Bureau of Navigation 1898*, 202.

20. Bernadou, "'Winslow' at Cardenas," 301–2.

21. Goodrich, "The St. Louis' Cable-Cutting," 158.

22. Winslow, "Cable-Cutting at Cienfuegos," 714. See also Evelyn M. Cherpak, "Cable Cutting at Cienfuegos," 119–22, 717. Winslow did not succeed in completely cutting telegraph communications, as a third uncut cable at Cienfuegos allowed communication to Jamaica and thence to Madrid; see Trask, *War with Spain*, 110.

23. Stockton, "Submarine Telegraph Cables in Time of War," 452–53. On exploration of the deep sea and its emergence as a sight of scientific study, see Rozwadowski, *Fathoming the Ocean*, particularly 79–95; on submarine telegraph cables, see Kennedy, "Imperial Cable Communications and Strategy," 728–52.

24. Squier, "The Influence of Submarine Telegraph Cables upon Military and Naval Supremacy," 621–22.

25. Halsey, "The Last Naval Engagement of the War," 53; U.S. War Department, *Military Notes on Cuba*, 251.

26. W. H. H. Southerland to Long, U.S.S. *Eagle*, Isle of Pines, Cuba, July 19, 1898, in *Appendix Bureau of Navigation 1898*, 247.

27. Ibid., 248.

28. Daniel Delehanty to Long, Navy Yard, Norfolk, September 1, 1898, in *Appendix Bureau of Navigation 1898*, 331–32. For other hydrographic problems associated with aiding Cuban insurgents, see W. T. Ryan to Long, U.S.S. *Peoria*, Key West, July 14, 1898, in *Appendix Bureau of Navigation 1898*, 690–91.

29. A. Marix to Long, U.S.S. *Scorpion*, off Manzanillo, July 1, 1898, in *Appendix Bureau of Navigation 1898*, 233; Halsey, "The Last Naval Engagement of the War," 55.

30. Halsey, "The Last Naval Engagement of the War," 55–57.

31. Ibid., 59–62.

32. Sampson, "The Atlantic Fleet in the Spanish War," 890–91.

33. Francis J. Higginson to William T. Sampson, At Sea, August 2, 1898, in *Appendix Bureau of Navigation 1898*, 636; ibid., 639.

34. Statement of Commander R. B. Bradford, October 14, 1898, 16, RG 43, Records of International Conferences, Commissions, and Expositions, Entry 800, Box 2, NARA; Linn, *The Philippine War*, 132.

35. R. W. Konter, "Wreck of the U.S.S. *Charleston*, Camiguin Island, Nov. 2nd, '99," May 1920, Charleston Operational File, Ships History Division, Naval History and Heritage Command, Washington, DC; Proceedings of a Court of Inquiry Convened to Investigate the Loss of the U.S.S. *Charleston*, Case 4931, RG 125, Records of the Office of the Judge Advocate General (Navy), Box 65, NARA.

36. *Albany* Court of Inquiry, 62.

37. Joseph E. Craig to George C. Remey, U.S.S. *Albany*, Cavite, P.I., December 21, 1900, *Albany* Court of Inquiry; ibid., 63.

38. Ibid., 23, 63.

39. Ibid., 76.

40. Tucker and Russell, *Natural Enemy, Natural Ally*, 5; Sampson, "The Atlantic Fleet in the Spanish War," 887; Commander J. E. Craig, 6th Endorsement, Bureau of Equipment, Hydrographic Office, July 13, 1898, RG 37, Records of the Hydrographic Office, Entry 32, Letters Sent and Received, February 1885–December 1901, Box 111, NARA.

41. Mahan, *The Influence of Sea Power upon History, 1660–1783*, 138.

42. U.S. Navy Department, *Annual Report of the Secretary of the Navy, 1903*, 365; Maclay, *A History of the United States Navy*, 239.


## Chapter Six

1. "At the Banquet Board," *Times*, December 29, 1898. Bradford was a member of the Sons of the American Revolution and also, as a descendant of William Bradford, the Society of Mayflower Descendants.

2. Morgan, *Making Peace with Spain*, 73.

3. Statement of Commander R. B. Bradford, October 14, 1898, 5, Record Group 43, Records of International Conferences, Commissions, and Expositions, Entry 800, Box 2, NARA, 7. See also McBride, *Technological Change in the United States Navy*, 90.

4. Statement of Commander R. B. Bradford, October 14, 1898, 14.

5. Coates et al., "Defending Nation, Defending Nature? Militarized Landscapes and Military Environmentalism in Britain, France, and the United States," 458. See also Pearson, Coates, and Cole, *Militarized Landscapes*.

6. U.S. Navy Department, Record of Proceedings of a Court of Inquiry in the Case of Rear-Admiral Winfield S. Schley, U.S. Navy, 1:140–43.

7. "How the Navy Appears in the Light of the Schley Inquiry," *Literary Digest*, October 19, 1901, Box 12, Folder 5, George Dewey Papers, MSS18366, Library of Congress, Washington, DC; Schley Court of Inquiry, 132.

8. Schley Court of Inquiry, 1768.

9. Captain F. E. Chadwick, Coal, Lecture Delivered 1901, RG 14, Faculty and Staff Presentations, 1901–1914, Box 2, Naval Historical Collection, Naval War College, Newport, RI; U.S. Navy Department, *Annual Report of the Secretary of the Navy, 1902*, 351; U.S. Navy Department, *Annual Report of the Secretary of the Navy, 1899*, 25.

10. Statement of Commander R. B. Bradford, October 14, 1898, 16; U.S. Navy Department, *Annual Report of the Secretary of the Navy, 1900*, 359; U.S. Navy Department, *Annual Report of the Secretary of the Navy, 1901*, 432.

11. Reidy, *Tides of History*, 294.

12. U.S. Navy Department, *Annual Report of the Secretary of the Navy*, 1902, 16; U.S. Navy Department, *Annual Report of the Secretary of the Navy, 1900*, 359; U.S. Navy Department, *Annual Report of the Secretary of the Navy for the Year 1901*, 432; M. Sicard, A. T. Mahan, and A. S. Crowninshield to John D. Long, Office of the Naval War Board, Washington, DC, 1898, RG 80, General Records of the Department of the Navy, General Board, Box 40, Folder 414-1, NARA; George Dewey to John D. Long, General Board, Newport, RI, August 23, 1901, RG 80, General Records of the Department of the Navy, General Board, Box 43, Folder 415, NARA; George Dewey, Relative to BLAKE Surveying the Locality of Dry Tortugas, 2nd Endorsement, General Board, Washington, DC, February 8, 1901, RG 80, General Records of the Department of the Navy, General Board, Box 43, Folder 415, NARA; C. H. Stockton to members of the United States Senate, in "Hydrographic Office," *United States Army and Navy Journal, and Gazette of the Regular and Volunteer Forces* 37 (1899–1900): 879. On Dewey's command of Narragansett, see Manning, *U.S. Coast Survey vs. Naval Hydrographic Office*, 30–31.

13. U.S. Navy Department, *Annual Report of the Secretary of the Navy, 1903*, 365; George Dewey to the Secretary of the Navy, General Board, Washington, D.C., October 10, 1900, RG 80, General Records of the Department of the Navy, General Board, Box 43, Folder 415, NARA; U.S. Navy Department, *Annual Report of the Secretary of the Navy, 1900*, 309; Manning, *U.S. Coast Survey vs. Naval Hydrographic*

*Office*, 135; Budget Appropriations can be found under "Hydrographic Office," in the report of the "Bureau of Equipment," U.S. Navy Department, *Annual Report of the Secretary of the Navy, 1898–1909.*

14. George L. Dyer to My Darling Darling Treasure, Sweetheart Wife, *Yankton*, Nipe Bay, March 25, 1901, Papers of George Leland Dyer, Manuscript Collection 340, Box 16, Folder B, Special Collections, Joyner Library, East Carolina University, Greenville, North Carolina; C. G. Calkins to Bradford, U.S.S. *Vixen*, Havana, Cuba, April 26, 1902, RG 19, Bureau of Equipment, General Correspondence, 1899–1910, Box 100, NARA; M. L. Wood to Bradford, U.S.S. *Eagle*, Port San Antonio, Jamaica, February 10, 1903, RG 19, Bureau of Equipment, General Correspondence, 1899–1910, Box 101, NARA. In this era, midshipmen were sometimes referred to as "naval cadets."

15. U.S. Navy Department, *Annual Report of the Secretary of the Navy, 1899*, 316, 318.

16. F. F. Fletcher to Leonard Wood, Havana, Cuba, March 27, 1900, RG 19, Bureau of Equipment, General Correspondence, 1899–1910, Box 104, NARA. See also LaFeber, *The New Empire*, and Karsten, *The Naval Aristocracy.*

17. Bradford to J. Aymer, Bureau of Equipment, Washington, DC, December 28, 1901, RG 19, Bureau of Equipment, General Correspondence, 1899–1910, Box 100, NARA; Knight to Bradford, U.S.S. *Yankton*, Santo Cruz del Sur, Cuba, February 6, 1903, RG 80, General Records of the Department of the Navy, General Board, Box 28, Folder 404-2, NARA; U.S. Navy Department, *Annual Report of the Secretary of the Navy, 1899*, 317; R. B. Bradford, 2nd Endorsement, January 20, 1903, RG 19, Bureau of Equipment, Correspondence, 1899–1910, Box 101, NARA.

18. U.S. Navy Department, *Annual Report of the Secretary of the Navy, 1900*, 358; Todd to J. M. Ellicott, Hydrographic Office, Washington, DC, November 24, 1900, RG 8, Intelligence and Technical Files, Box 36, Folder 1, Naval Historical Collection, Naval War College (NWC).

19. H.R., 57th Cong., 1st Sess., Doc. No. 140, Establishment of Naval Stations in the Philippine Islands, at 2–3, (1901).

20. Ibid., 22. On Remey's service in command of the Asiatic Squadron and, quite superficially, as president of the Remey Board, see Remey, *Life and Letters of Rear Admiral George Collier Remey*, 859–918.

21. Lieutenant J. M. Ellicott, "The Strategic Features of the Philippine Islands, Hawaii, and Guam," 1900, RG 8, Intelligence and Technical Files, Box 104, Folder 6, Naval Historical Collection, NWC; Rodgers to Long, U.S.S. *New York*, Olongapo, P.I., June 12, 1901, RG 45, Office of Naval Records and Library, Subject File PS, Philippine Islands, Box 563, NARA. See also U.S. Navy Department, *Annual Report of the Secretary of the Navy, 1901*, 370–71.

22. See Manning, *U.S. Coast Survey vs. Naval Hydrographic Office*, 129–51.

23. C. C. Todd to Cameron McRae Winslow, Hydrographic Office, Washington, DC, April 25, 1900, RG80, General Records of the Navy Department, General

Correspondence, 1897–1915, Box 486, NARA; Royal B. Bradford, "Memorandum for the Secretary of the Navy in Relation to Ocean and Lake Surveys," RG 80, General Records of the Navy Department, General Correspondence, 1897–1915, Box 487, NARA; *Congressional Record*, 56th Cong., April 19, at 4442 (1900).

24. *Congressional Record*, 56th Cong., April 19, at 4428 (1900); "Statement of Mr. Henry S. Pritchett," Naval Affairs Committee, March 19, 1900.

25. *Congressional Record*, 56th Cong., June 7, at 6881 (1900).

26. Ibid., 6884.

27. Ellicott, "The Strategic Features of the Philippine Islands, Hawaii, and Guam," Naval Historical Collection, NWC; Knight to Bradford, U.S.S. *Yankton*, Portsmouth, NH, July 8, 1903, RG 19, Bureau of Equipment, General Correspondence, 1899–1910, Box 101, NARA; U.S. Navy Department, *Annual Report of the Secretary of the Navy, 1901*, 75; ibid., 18, 371.

28. See Hattendorf, Simpson, and Wadleigh, *Sailors and Scholars*, 38–45, and Spector, *Professors of War*, 64–100.

29. Lieut-Cmdr Selfridge and Lieut Ellicott, "Reconnaissance of Fisher's Island Sound, Gardiner's and Peconic Bays, Shinnecock Canal, etc.," August 27, 1896, RG 8, Intelligence and Technical Files, Box 36, Folder 13, Naval Historical Collection, NWC.

30. Commander C. H. Stockton, "The Formation of Maps or Charts for War or Coast Defense Purposes," 1899, Manuscript Collection 56, Papers of Charles H. Stockton, Box 3, Naval Historical Collection, NWC; Captain H. C. Taylor, Closing Address, Session of 1895, RG 28, President's File, Henry Clay Taylor, Naval Historical Collection, NWC; Naval War College, *Abstract of Course, 1895*, 27–28.

31. See Hattendorf, Simpson, and Wadleigh, *Sailors and Scholars*, 40–41, 56; Spector, *Professors of War*, 74–87; and McCarty-Little, "The Strategic Naval War Game or Chart Maneuver," 1213–33.

32. McCarty Little, "The Strategic Naval War Game," 1213, 1219.

33. Ibid., 1220.

34. Commander A. M. Knight to Bradford, U.S.S. *Yankton*, Santa Cruz del Sur, Cuba, February 6, 1903, RG 80, General Records of the Department of the Navy, General Board, Box 28, Folder 404-2, NARA; Knight to Bradford, U.S.S. *Yankton*, Cienfuegos, Cuba, June 6, 1902, RG 19, Bureau of Equipment, General Correspondence, 1899–1910, Box 100, NARA; George Dewey, 3rd Endorsement, June 20, 1902, RG 80, General Records of the Department of the Navy, General Board, Box 28, Folder 404-2, NARA.

35. *Washington Times*, June 15, 1902, Dewey Papers, Box 12, Folder 3, LOC. On the Venezuelan Crisis and the Winter Maneuvers, see Livermore, "Theodore Roosevelt, the American Navy, and the Venezuelan Crisis of 1902–03," 452–71; Spector, *Admiral of the New Empire*, 137–53.

36. *Washington Evening Times*, October 17, 1902, Dewey Papers, Box 12, Folder 3, LOC.

37. Journal of the Commander in Chief, January 3, 1903, Dewey Papers, Box 44, Folder 3, LOC; Henry C. Taylor, memorandum, January 3, 1902, Dewey Papers, Box 44, Folder 3, LOC.

38. Journal of the Commander in Chief, January 19, 1903, Dewey Papers, Box 44, Folder 3, LOC.

39. See Spector, *Admiral of the New Empire*, 154–78; Challener, *Admirals, Generals, and American Foreign Policy*, 36.

40. G. L. Dyer to Bradford, U.S.S. *Yankton*, Santiago de Cuba, May 17, 1899, RG 80, General Records of the Department of the Navy, General Board, Box 28, Folder 404-2; Fletcher to Bradford, U.S.S. *Eagle*, Caimanera, Cuba, May 25, 1899, RG 80, General Records of the Department of the Navy, General Board, Box 28, Folder 404-2; McCalla, 12th Endorsement, Navy Yard, Norfolk, Virginia, June 29, 1899, RG 80, General Records of the Department of the Navy, General Board, Box 28, Folder 404-2; General Board, Newport, RI, August 21, 1901, M1493, Proceedings and Hearings of the General Board of the U.S. Navy, 1900–1950, Roll 1, NARA.

41. R. D. Evans to Long, March 25, 1902, RG 80, General Records of the Department of the Navy, General Board, Box 28, Folder 404-2.

42. Dewey to William H. Moody, June 16, 1903, RG 80, General Records of the Department of the Navy, General Board, Box 28, Folder 405, NARA; ibid., June 8, 1903; H. C. Taylor to Long, General Board, Washington, DC, September 26, 1901, RG 80, General Records of the Department of the Navy, General Board, Box 28, Folder 405, NARA. On Subig in the larger strategic context of naval planning in the twentieth century, see Miller, *War Plan Orange*, 65–76.

43. Daniel Joseph Costello, "Planning for War"; quotes taken from Costello, n.p.; on the problems associated with the defense of Subic Bay, see General Board, Washington, DC, November 25, 1901, M1493, Proceedings and Hearings of the General Board of the U.S. Navy, 1900–1950, Roll 1, NARA; ibid., April 25, 1902; General Board, Newport, Rhode Island, August 27, 1902, M1493, Proceedings and Hearings of the General Board of the U.S. Navy, 1900–1950, Roll 1, NARA; See Challener, *Admirals, Generals, and American Foreign Policy*, 39–40, 193, 228; Livermore, "American Naval-Base Policy in the Far East," 130–32; Braisted, "The United States Navy's Dilemma in the Pacific, 1906–1909," 235–44; Melville, "Important Elements in Naval Conflicts," 130–31; Braisted, *The United States Navy in the Pacific*, 121–24; Challener, *Admirals, Generals, and American Foreign Policy*, 47–50, 233–41; on Bradford's squabble with Taylor, see *The Times* (Washington, DC), July 20, 1902, 9.

44. Braisted, *The United States Navy in the Pacific*, 124–36; Livermore, "American Naval-Base Policy in the Far East," 122–25.

45. Statement of Henry S. Pritchett, "Efforts Made by the Navy Department to Obtain Control of the United States Coast and Geodetic Survey, April 1900," RG 23, Records of The Coast and Geodetic Survey, "Superintendent's File," 1866–1910, Box 529.

46. "Statement of Rear Admiral R. B. Bradford," Hearings before the Naval Affairs Committee, December 6, 1902, Congressional Record.

47. Charles D. Sigsbee to Charles H. Allen, Office of Naval Intelligence, Washington, DC, March 20, 1901, RG 80, General Records of the Department of the Navy, General Board, Box 28, Folder 404-3, NARA; Mahan, "The Advantages of Subig Bay over Manila as a Base in the Philippine Islands," in Seager and Maguire, *Letters and Papers of Alfred Thayer Mahan*, 3:659; Captain H. C. Taylor, "Address Delivered before the First Naval Battalion, N.Y., Wednesday, January 9, 1895," RG 28, Henry Clay Taylor President's File, Naval Historical Collection, NWC; Frederick Rodgers to John D. Long, Olongapo, P.I., June 12, 1901, RG 45, Office of Naval Records and Library, Subject File PS, Philippine Islands, Box 563, NARA; Captain Asa Walker, "Notes on Cuban Ports," RG 8, Intelligence and Technical Files, Box 70, Folder 1, Naval Historical Collection, NWC; George Dewey to John D. Long, General Board, Washington, DC, April 23, 1901, RG 80, General Records of the Department of the Navy, General Board, Box 38, Folder 413, NARA.

## Epilogue

1. Both biblical verses appear below stained glass in the Chapel of the United States Naval Academy.

2. Steinbeck's account can be found in "Appendix: About Ed Ricketts," in Steinbeck, *The Log from the Sea of Cortez*, 267–71.

3. Steinbeck to Frank Knox, Palisades, California, May 5, 1942, in Elaine Steinbeck and Robert Wallsten, *Steinbeck*, 246–47; Steinbeck, *The Log from the Sea of Cortez*, 269–71; Morison, *History of United States Naval Operations in World War II*, 7:151–54, 182–83. Tarawa was, in fact, not a Japanese mandate, but a British possession prior to the Japanese occupation. There is no record of this meeting in the correspondence of RG 38, Records of the Office of Naval Intelligence, NARA, College Park, MD.

4. See Weir, *An Ocean in Common*; Hamblin, *Poison in the Well*; and Hamblin, *Arming Mother Nature*. For an excellent analysis of the Mekong environment's role in the imperial state making, see Biggs, *Quagmire*. The navy's continuing tradition of brown water operations—Mahan aside—is the central argument in Hagan, *This People's Navy*.

5. Bates, Tselepis, and Von Nieda, "Needed: Shallow Thinking," 43–51; Crowninshield to John D. Long, Washington, DC, March 7, 1900, in Allen, *Papers of John Davis Long*, 318; U.S. Navy Department, *Annual Report of the Secretary of the Navy, 1905*, 325.

6. B.A. Friedman, *21ˢᵗ Century Ellis: Operational Art and Strategic Prophecy for the Modern Era* (Annapolis: Naval Institute Press, 2015), 87–88; E.H. Ellis, "Naval Bases, their Location and Resources," 2, Record Group 8, Box 80, Naval Historical Collection, NWC. On Ellis, see also Dirk Anthony Ballendorf and Merrill Lewis Bartlett, *Pete Ellis: An Amphibious Warfare Prophet, 1880–1923* (Annapolis: Naval Institute Press, 1997).

7. United States Department of Defense, "Quadrennial Defense Review" (February 2010), 84–88.

8. Sponsel, "Coral Reefs as Objects of Scientific Study, from Threatening to Threatened," (paper presented at the Annual Meeting of the History of Science Society, Chicago, November 2014); Sponsel, "From Cook to Cousteau: The Many Lives of Coral Reefs," in Gillis and Torma, *Fluid Frontiers*, 137–61; Rozwadowski, *Fathoming the Ocean*, v.

## Acknowledgments

1. Diary of Robert Weir, *Clara Bell*, Log 164, May 5, 1858, Collections Research Center, Mystic Seaport Museum, Mystic, CT.

# Bibliography

## Manuscript Collections

Albany, New York
  New York Historical Society
    Charles D. Sigsbee Papers
Annapolis, Maryland
  Special Collections and Archives, Nimitz Library, U.S. Naval Academy
Greenville, North Carolina
  Special Collections, Joyner Library, East Carolina University
    George Leland Dyer Papers
Lancaster, Pennsylvania
  Archives and Special Collections, Franklin and Marshall College
    Reynolds Family Papers
Mystic, Connecticut
  Collections Research Center (CRC). Mystic Seaport Museum (MSM)
    Diary of Humphrey Hill, *Magdala*, Log 908
    Diary of Nelson Cole Haley, *Charles W. Morgan*, Log 145
    Diary of Reverend Thomas Douglass, *Morrison*, Log 343
    Diary of Robert Weir, *Clara Bell*, Log 164
    Diary of Theodore Lewis, *Atlantic*, Log 822
    Diary of Unknown Author, *Weymouth/Indramayoe*, Log 507
New Bedford, Massachusetts
  New Bedford Whaling Museum Research Library and Archives
    Diary of William Abbe, *Atkins Adams*, Log 485
New Haven, Connecticut
  Beinecke Rare Book and Manuscript Library, Yale University
    Journal of George Foster Emmons, WA MSS 166
    Journal of George M. Colvocoresses, WA MSS 101
    Journal of Silas Holmes, WA MSS 260
Newport, Rhode Island
  Naval Historical Collection, Naval War College (NWC)
    Papers of Charles H. Stockton, Manuscript Collection 56
    Record Group (RG) 8, Intelligence and Technical Files
    Record Group 14, Faculty and Staff Presentations
    Record Group 28, Presidents' File

Washington, DC
    Library of Congress (LOC), Manuscript Division
        Papers of George Dewey
        Papers of George E. Belknap
        Papers of Matthew Fontaine Maury
    Library of Congress, Geography and Map Division
    National Archives and Records Administration (NARA)
        Cartographic and Architectural Section
        Journal of Frederick D. Stuart, NA 313, Roll 20
        Journal of George T. Sinclair, NA 314, Roll 21
        Proceedings and Hearings of the General Board of the U.S. Navy,
            1900–1950, M1493
        Record Group 19, Records of the Bureau of Ships
        Record Group 23, Records of the Coast and Geodetic Survey
        Record Group 27, Records of the Weather Bureau
        Record Group 37, Records of the Hydrographic Office
        Record Group 38, Records of the Office of Naval Intelligence
        Record Group 43, Records of International Conferences, Commissions,
            and Expositions
        Record Group 45, Office of Naval Records and Library
        Record Group 78, Records of the Naval Observatory
        Record Group 80, General Records of the Department of the Navy
        Record Group 125, Records of the Office of the Judge Advocate General
            (U.S. Navy)
        Records Relating to the U.S. Exploring Expedition
    Naval History and Heritage Command
        USS *Charleston* Operational File

## Government Documents

Baker, Marcus. *Geographic Dictionary of Alaska.* Washington, DC: Government
    Printing Office, 1906.
Belknap, George E. *Deep-Sea Soundings in the North Pacific Ocean, Obtained in the
    United States Steamer* Tuscarora. *Congressional Record.* 56th Cong. 1st Sess.
    Washington, DC: Government Printing Office, 1874.
Davis, Charles H. *Narrative of the North Polar Expedition.* Washington, DC:
    Government Printing Office, 1876.
*Hearing before the Subcommittee of House Committee on Appropriations, Sundry
    Civil Appropriations Bill for 1900.* 55th Cong. Washington, DC: Government
    Printing Office, 1899.
Hughes, W. S. *Founding and Development of the U.S. Hydrographic Office.*
    Washington, DC: Government Printing Office, 1887.

Morris, R. O. *Charts and Surveys in Peace and War: The History of the RN Hydrographic Service, 1919–1970*. London: Her Majesty's Stationary Office, 1995.

Sigsbee, Charles D. *Deep-Sea Sounding and Dredging: A Description and Discussion of the Methods and Appliances Used on Board the Coast and Geodetic Survey Steamer,* Blake. Washington, DC: Government Printing Office, 1880.

U.S. Congress. 32nd Cong., 1st Sess. *Congressional Globe.*

U.S. Congress. 40th Cong., 1st Sess. *Congressional Globe.*

U.S. Congress. 56th Cong., 1st Sess. *Congressional Record.*

U.S. Department of Defense. *Quadrennial Defense Review,* February 2010.

U.S. House of Representatives. 42nd Cong. 3rd Sess. *Executive Documents.*

U.S. House of Representatives. 57th Cong. 1st Sess. Document No. 140, Establishment of Naval Stations in the Philippine Islands.

U.S. Naval War College. *Abstract of Course, 1895.* Washington, DC: Government Printing Office, 1895.

U.S. Navy Department. *Annual Report of the Navy Department for the Year 1898: Appendix to the Report of the Chief of the Bureau of Navigation.* Washington, DC: Government Printing Office, 1898.

———. *Annual Report of the Secretary of the Navy.* Washington, DC: Government Printing Office, 1830, 1868–71, 1873, 1874, 1878, 1883, 1885, 1898–1903, 1905–1909.

———. Record of Proceedings of a Court of Inquiry in the Case of Rear-Admiral Winfield S. Schley, U.S. Navy. 2 vols. Washington, DC: Government Printing Office, 1902.

———. *Rules, Regulations, and Instructions, for the Naval Service of the United States.* Washington, DC: E. De Krafft, 1818.

U.S. Senate. 53rd Cong., 3rd Sess. *Executive Documents.*

U.S. War Department. *Military Notes on Cuba.* Washington, DC: Government Printing Office, 1898.


## Books

Agassiz, Alexander. *Three Cruises of the United States Coast and Geodetic Survey Steamer* Blake: *In the Gulf of Mexico, in the Caribbean Sea, and along the Atlantic Coast, from 1877 to 1880.* New York: Houghton, Mifflin, 1888.

Akerman, James R., ed. *The Imperial Map: Cartography and the Master of Empire.* Chicago: University of Chicago Press, 2009.

Allen, Gardner Weld, ed. *The Papers of John Davis Long, 1897–1904.* Boston: Massachusetts Historical Society, 1939.

Anderson, Benedict. *Imagined Communities: Reflections on the Origin and Spread of Nationalism.* Rev. ed. New York: Verso, 1991.

Andersson, Axel. *Hero of the Atomic Age: Thor Heyerdahl and the Kon-Tiki Expedition.* Oxford: Peter Lang, 2010.

Andrewes, William J. H., ed. *The Quest for Longitude: The Proceedings of the Longitude Symposium, Harvard University, Cambridge, Massachusetts, November 4–6, 1993*. Cambridge, MA: Collection of Historical Scientific Instruments, 1996.

Arens, W. *The Man-Eating Myth: Anthropology and Anthropophagy*. New York: Oxford University Press, 1979.

Army and Navy Chronicle. Vol. 4. Washington, DC: B. Homans, 1837.

Baker, Anne. *Heartless Immensity: Literature, Culture, and Geography in Antebellum America*. Ann Arbor: University of Michigan Press, 2006.

Baker, Margaret. *Folklore of the Sea*. London: David & Charles, 1979.

Ballendorf, Dirk Anthony, and Merrill Lewis Bartlett. *Pete Ellis: An Amphibious Warfare Prophet, 1880–1923*. Annapolis: Naval Institute Press, 1997.

Barker, Francis, Peter Hulme, and Margaret Iverson, eds. *Cannibalism and the Colonial World*. New York: Cambridge University Press, 1998.

Bates, Charles C. *HYDRO to NAVOCEANO: 175 Years of Ocean Survey and Prediction by the U.S. Navy, 1830–2005*. Edited by George L. Hanssen. Rockton, IL: Cornfield Press, 2005.

Bauer, K. Jack. *Surfboats and Horse Marines: U.S. Naval Operations in the Mexican War, 1846–48*. Annapolis: Naval Institution Press, 1969.

Beck, Horace. *Folklore and the Sea*. Mystic, CT: The Marine Historical Association, Incorporated, 1973.

Bercaw Edwards, Mary K. *Cannibal Old Me: Spoken Sources in Melville's Early Works*. Kent, OH: Kent State University Press, 2009.

Biggs, David. *Quagmire: Nation-Building and Nature in the Mekong Delta*. Seattle: University of Washington Press, 2010.

Blum, Hester. *The View from the Masthead: Maritime Imagination and Antebellum American Sea Narratives*. Chapel Hill: University of North Carolina Press, 2008.

Blunt, Edmund M. *The American Coast Pilot: Containing Directions for the Principal Harbours, Capes, and Headlands, on the Coast of North and South America*. 12th ed. New York: Edmund and George W. Blunt, 1833.

Bolles, Blair. *Tyrant from Illinois: Uncle Joe Cannon's Experiment with Personal Power*. New York: W.W. Norton, 1951.

Bolster, W. Jeffrey. *Black Jacks: African American Seamen in the Age of Sail*. Cambridge, MA: Harvard University Press, 1997.

———. *The Mortal Sea: Fishing the Atlantic in the Age of Sail*. Cambridge, MA: Harvard University Press, 2012.

Boot, Max. *The Savage Wars of Peace: Small Wars and the Rise of American Power*. New York: Basic Books, 2002.

Bowditch, Nathaniel. *The New American Practical Navigator: Being an Epitome of Navigation*. New York: Edmund M. Blunt, 1826.

Bradford, James C., ed. *Captains of the Old Steam Navy: Makers of the American Naval Tradition, 1840–1880*. Annapolis, MD: Naval Institute Press, 1986.

Bradford, William. *Of Plymouth Plantation, 1620–1647: The Complete Text,* ed. Samuel Eliot Morison. New York: Alfred A. Knopf, 1952.

Brady, Lisa. *War upon the Land: Military Strategy and the Transformation of Southern Landscapes during the American Civil War.* Athens: University of Georgia Press, 2012.

Braisted, William R. *The United States Navy in the Pacific, 1897–1909.* Austin: University of Texas Press, 1958.

Brewster, Mary. *"She Was a Sister Sailor": The Whaling Journals of Mary Brewster.* Edited by Joan Druett. Mystic, CT: Mystic Seaport Museum, 1992.

Brooke, George M., Jr. *John Mercer Brooke: Naval Scientist and Educator.* Charlottesville: University Press of Virginia, 1980.

Brooke, John M. *John M. Brooke's Pacific Cruise and Japanese Adventure, 1858–1860.* Edited by George M. Brooke Jr. Honolulu: University of Hawaii Press, 1986.

Browne, J. Ross. *Etchings of a Whaling Cruise.* Edited by John Seelye. Cambridge, MA: Harvard University Press, 1968.

Bruce, Robert V. *The Launching of Modern American Science, 1846–1876.* New York: Alfred A. Knopf, 1987.

Brückner, Martin. *The Geographic Revolution in Early America: Maps, Literacy, and National Identity.* Chapel Hill: University of North Carolina Press, 2007.

Burnett, D. Graham. *Masters of All They Surveyed: Exploration, Geography, and a British El Dorado.* Chicago: University of Chicago Press, 2000.

———. *Trying Leviathan: The Nineteenth-Century New York Court Case That Put the Whale on Trial and Challenged the Order of Nature.* Princeton, NJ: Princeton University Press, 2007.

Burstyn, Harold L. *At the Sign of the Quadrant: An Account of the Contributions to American Hydrography Made by Edmund March Blunt and His Sons.* Mystic, CT: The Marine Historical Association, 1957.

Chadwick, French E. *The Relations of the United States and Spain: The Spanish-American War.* 2 vols. New York: Charles Scribner's, 1911.

Challener, Richard D. *Admirals, Generals, and American Foreign Policy, 1898–1914.* Princeton, NJ: Princeton University Press, 1973.

Chase, Owen. *Narrative of the Most Extraordinary and Distressing Shipwreck of the Whaleship* Essex: *With a Supplementary Account of Survivors and Herman Melville's Memoranda on Owen Chase.* New York: Lyons Press, 1999.

Cleaver, Anne Hoffman, and E. Jeffrey Stann, eds. *Voyage to the Southern Ocean: The Letters of Lieutenant William Reynolds from the U.S. Exploring Expedition, 1838–1842.* Annapolis, MD: Naval Institute Press, 1988.

Cloud, John. *Science on the Edge: The Story of the U.S. Coast and Geodetic Survey, Its Transition into ESSA and NOAA, and the American Triumph of the Earth Sciences in the 20th Century.* National Oceanic and Atmospheric Administration. http://www.lib.noaa.gov/noaainfo/heritage/coastandgeodeticsurvey/Pritchettchapter.pdf. Accessed February 10, 2012.

Cochelet, Charles. *Narrative of the Shipwreck of the Sophia*. London: Sir Richard Phillips, 1822.

Cooper, James Fenimore. *History of the Navy of the United States of America*. Vol. 1. 3rd ed. Cooperstown, NY: H & E Phinney, 1847.

———. *The Pilot: A Tale of the Sea*. New York: J. G. Gregory, 1862.

———. *The Prairie: A Tale*. New York: W.A. Townsend, 1859.

———. *The Red Rover: A Tale*. London: Richard Bentley, 1836.

Creighton, Margaret S. *Rites and Passages: The Experience of American Whaling, 1830–1870*. New York: Cambridge University Press, 1995.

Cronon, William, ed. *Uncommon Ground: Toward Reinventing Nature*. New York: W.W. Norton, 1995.

Cumings, Bruce. *Dominion from Sea to Sea: Pacific Ascendancy and American Power*. New Haven, CT: Yale University Press, 2009.

Cummings, Henry. *A Synopsis of the Cruise of the U.S.S. "Tuscarora."* San Francisco: Cosmopolitan Steam, 1874.

Dana, Richard Henry. *Two Years before the Mast*. In *Two Years Before the Mast, and Other Voyages*. New York: Library of America, 2005.

———. *Two Years before the Mast, and Other Voyages*. New York: Library of America, 2005.

DeArmond, Robert N. *The USS Saginaw in Alaska Waters, 1867–1868*. Edited by Richard A. Pierce. Fairbanks, AK: Limestone Press, 1997.

Dening, Greg. *Islands and Beaches: Discourse on a Silent Land, Marquesas, 1774–1880*. Honolulu: University Press of Hawaii, 1980.

Dick, Steven J. *Sky and Ocean Joined: The U.S. Naval Observatory, 1830–2000*. New York: Cambridge University Press, 2003.

Dolan, Eric Jay. *Leviathan: The History of Whaling in America*. New York: W.W. Norton, 2007.

Drake, Brian Allan. *The Blue, the Gray, and the Green: Toward an Environmental History of the Civil War*. Athens, GA: University of Georgia Press, 2015.

Drake, Frederick C. *The Empire of the Seas: A Biography of Rear Admiral Robert Wilson Shufeldt, USN*. Honolulu: University of Hawaii Press, 1984.

Driver, Felix, and Luciana Martins, eds. *Tropical Visions in the Age of Empire*. Chicago: University of Chicago Press, 2005.

Dudley, William S., and Michael Crawford, eds. *The Early Republic and the Sea: Essays on the Naval and Maritime History of the Early United States*. Washington, DC: Brassey's, 2001.

Dupree, A. Hunter. *Science in the Federal Government: A History of Policies and Activities to 1940*. Cambridge, MA: Harvard University Press, 1957.

Edney, Matthew H. *Mapping an Empire: The Geographical Construction of British India, 1765–1843*. Chicago: University of Chicago Press, 1997.

Endicott, William. *Wrecked among Cannibals in the Fijis: A Narrative of Shipwreck and Adventure in the South Seas*. Salem, MA: Marine Research Society, 1923.

Fabian, Ann. *The Skull Collectors: Race, Science, and America's Unburied Dead.* Chicago: University of Chicago Press, 2010.

Ferguson, Eugene S. *Truxtun of the Constellation: The Life of Commodore Thomas Truxtun, U.S. Navy, 1755–1822.* 2nd ed. Baltimore, MD: Johns Hopkins University Press, 2000.

Fleming, James Rodger. *Meteorology in America, 1800–1870.* Baltimore: Johns Hopkins University Press, 1990.

Friedman, B.A. *21st Century Ellis: Operational Art and Strategic Prophecy for the Modern Era.* Annapolis: Naval Institute Press, 2015.

Gilje, Paul A. *Liberty on the Waterfront: American Maritime Culture in the Age of Revolution.* Philadelphia: University of Pennsylvania Press, 2007.

Gillis, John, and Franziska Torma. *Fluid Frontiers: New Currents in Marine Environmental History.* Cambridge, U.K.: White Horse Press, 2015.

Goetzmann, William. *Exploration and Empire: The Explorer and the Scientist in the Winning of the American West.* New York: Knopf, 1966.

——. *New Lands, New Men: America and the Second Great Age of Discovery.* New York: Viking, 1986.

Grady, John. *Matthew Fontaine Maury: Father of Oceanography.* Jefferson, NC: McFarland & Co., 2015.

Greenberg, Amy S. *Manifest Manhood and the Antebellum American Empire.* New York: Cambridge University Press, 2005.

Guttridge, Leonard F. *Icebound: The Jeanette Expedition's Quest for the North Pole.* Annapolis, MD: Naval Institute Press, 1986.

Hagan, Kenneth J. *American Gunboat Diplomacy and the Old Navy, 1877–1889.* Westport, CT: Greenwood Press, 1973.

——, ed. *In Peace and War: Interpretations of American Naval History, 1775–1984.* Westport, CT: Greenwood Press, 1984.

——. *This People's Navy: The Making of American Sea Power.* New York: Free Press, 1991.

Hamblin, Jacob Darwin. *Arming Mother Nature: The Birth of Catastrophic Environmentalism.* New York: Oxford University Press, 2013.

——. *Poison in the Well: Radioactive Waste in the Oceans at the Dawn of the Nuclear Age.* New Brunswick, NJ: Rutgers University Press, 2008.

Harley, J. B., and David Woodward, eds. *Cartography in Prehistoric, Ancient, and Medieval Europe and the Mediterranean.* Vol. 1 of *The History of Cartography.* Chicago: University of Chicago Press, 1987.

Hattendorf, John B., ed. *Doing Naval History: Essays Toward Improvement.* Newport, RI: Naval War College Press, 1995.

——, ed. *The Influence of History on Mahan: The Proceedings of a Conference Marking the Centenary of Alfred Thayer Mahan's "The Influence of Sea Power upon History, 1660–1783."* Newport, RI: Naval War College Press, 1991.

——, ed. *Mahan on Naval Strategy: Selections from the Writings of Rear Admiral Alfred Thayer Mahan.* Annapolis: Naval Institute Press, 1991.

——, ed. *The Oxford Encyclopedia of Maritime History*. 4 vols. New York: Oxford University Press, 2007.

——, ed. *Ubi Sumus? The State of Naval and Maritime History*. Newport, RI: Naval War College Press, 1994.

Hattendorf, John B., Mitchell Simpson III, and John R. Wadleigh. *Sailors and Scholars: The Centennial History of the U.S. Naval War College*. Newport, RI: Naval War College Press, 1984.

Headrick, Daniel R. *The Tentacles of Progress: Technology Transfer in the Age of Imperialism, 1850–1940*. New York: Oxford University Press, 1988.

——. *The Tools of Empire: Technology and European Imperialism in the Nineteenth Century*. New York: Oxford University Press, 1981.

Hecht, Susanna B. *The Scramble for the Amazon and the "Lost Paradise" of Euclides da Cunha*. Chicago: University of Chicago Press, 2013.

Heflin, Wilson L. *Herman Melville's Whaling Years*. Edited by Mary K. Bercaw Edwards and Thomas Feral Heffernan. Nashville, TN: Vanderbilt University Press, 2004.

Herman, Jan K. *A Hilltop in Foggy Bottom: Home of the Old Naval Observatory and the Navy Medical Department*. Washington, DC: Naval Medical Command, 1984.

Herrick, Walter R. *The American Naval Revolution*. Baton Rouge: Louisiana State University Press, 1967.

Herwig, Holger. *The Politics of Frustration: The United States in German Naval Planning, 1889–1941*. Boston: Little, Brown, 1976.

Hoganson, Kristin L. *Fighting for American Manhood: How Gender Politics Provoked the Spanish-American and Philippine-American Wars*. New Haven, CT: Yale University Press, 1998.

Hornsby, Stephen J. *Surveyors of Empire: Samuel Holland, J.F.W. des Barres and the Making of the Atlantic Neptune*. Montreal: McGill-Queen's University Press, 2011.

Howse, Derek. *Greenwich Time and the Discovery of the Longitude*. New York: Oxford University Press, 1990.

Igler, David. *The Great Ocean: Pacific Worlds from Captain Cook to the Gold Rush*. New York: Oxford University Press, 2013.

Jacob, Christian. *The Sovereign Map: Theoretical Approaches in Cartography throughout History*. Translated by Tom Conley. Chicago: University of Chicago Press, 2005.

Jampoler, Andrew C. A. *Sailors in the Holy Land: The 1848 American Expedition to the Dead Sea and the Search for Sodom and Gomorrah*. Annapolis, MD: Naval Institute Press, 2005.

Jomini, Baron Antoine-Henri de. *The Art of War*. Philadelphia: J.B. Lippincott, 1862.

Jones, Ryan T. *Empire of Extinction: Russians and the North Pacific's Strange Beasts of the Sea, 1741–1867*. New York: Oxford University Press, 2014.

Joyce, Barry Alan. *The Shaping of American Ethnography: The Wilkes Exploring Expedition, 1838–1842*. Lincoln: University of Nebraska Press, 2001.

Judd, Richard W. *Second Nature: An Environmental History of New England*. Amherst: University of Massachusetts Press, 2014.

Karsten, Peter. *The Naval Aristocracy: The Golden Age of Annapolis and the Emergence of Modern American Navalism*. New York: Free Press, 1972.

Kasson, John F. *Civilizing the Machine: Technology and Republican Values in America, 1776–1900*. New York: Grossman, 1976.

Keiner, Christine. *The Oyster Question: Scientists, Watermen, and the Maryland Chesapeake Bay since 1880*. Athens: University of Georgia Press, 2009.

Kroll, Gary. *America's Ocean Wilderness: A Cultural History of Twentieth-Century Exploration*. Lawrence: University of Kansas Press, 2008.

Kuhn, Thomas S. *The Structure of Scientific Revolutions*. Chicago: University of Chicago Press, 1962.

Labaree, Benjamin. *The Atlantic World of Robert G. Albion*. Middletown, CT: Wesleyan University Press, 1975.

Labaree, Benjamin W., William M. Fowler Jr., John B. Hattendorf, Jeffrey J. Safford, Edward W. Sloan, and Andrew W. German. *America and the Sea: A Maritime History*. Mystic, CT: Mystic Seaport, 1998.

LaFeber, Walter. *The New Empire: An Interpretation of American Expansion, 1860–1898*. Ithaca: Cornell University Press, 1963.

Ledyard, John. *The Last Voyage of Captain Cook: The Collected Writings of John Ledyard*. Edited by James Zug. Washington, DC: National Geographic Society, 2005.

Leeman, William P. *The Long Road to Annapolis: The Founding of the Naval Academy and the Emerging American Republic*. Chapel Hill: University of North Carolina Press, 2010.

Lewis, Charles Lee. *Matthew Fontaine Maury: The Pathfinder of the Seas*. Annapolis, MD: Naval Institute Press, 1927.

Lewis, Michael, ed. *American Wilderness: A New History*. New York: Oxford University Press, 2007.

Lightman, Bernard, ed. *Victorian Science in Context*. Chicago: University of Chicago Press, 1997.

Lindberg, David C., and Ronald L. Numbers, eds. *God and Nature: Historical Essays on the Encounter between Christianity and Science*. Berkeley: University of California Press, 1986.

Linn, Brian McAllister. *The Philippine War: 1899–1902*. Lawrence: Kansas University Press, 2000.

Lipman, Andrew. *The Saltwater Frontier: Indians and the Contest for the American Coast*. New Haven, CT: Yale University Press, 2015.

Livezey, William E. *Mahan on Sea Power*. Norman: University of Oklahoma Press, 1947.

Long, John Davis. *The New American Navy*. 2 vols. New York: Outlook, 1903.

Love, Robert W., Jr. *History of the U.S. Navy.* Vol. 1. Harrisburg: Stackpole Books, 1992.

Lyons, Paul. *American Pacificism: Oceania in the U.S. Imagination.* New York: Routledge, 2006.

Maclay, Edgar Stanton. *A History of the United States Navy: From 1775 to 1902.* 3 vols. New York: D. Appleton, 1902.

Mahan, Alfred T. *The Influence of Sea Power upon History, 1660–1783.* 5th edition. Boston: Little, Brown, 1894.

Manning, Charles G. *U.S. Coast Survey vs. Naval Hydrographic Office: A 19th-Century Rivalry in Science and Politics.* Tuscaloosa: University of Alabama Press, 1988.

Marx, Leo. *The Machine in the Garden: Technology and the Pastoral Ideal in America.* New York: Oxford University Press, 1964.

Maury, Matthew Fontaine. *Explanations and Sailing Directions to Accompany the Wind and Current Charts.* Vol. 1. 3rd ed. Washington, DC: C. Alexander, 1851.

———. *Explanations and Sailing Directions to Accompany the Wind and Current Charts.* Vol. 1. 8th ed. Washington, DC: William A. Harris, 1858.

———. *A New Theoretical and Practical Treatise on Navigation.* 3rd ed. Philadelphia: E.C. & J. Biddle, 1845.

———. *The Physical Geography of the Sea.* New York: Harper, 1855.

McBride, William M. *Technological Change and the United States Navy, 1865–1945.* Baltimore, MD: Johns Hopkins University Press, 2000.

McEvoy, Arthur F. *The Fishermen's Problem: Ecology and Law in the California Fisheries, 1850–1980.* New York: Cambridge University Press, 1986.

McKee, Christopher. *A Gentlemanly and Honorable Profession: The Creation of the U.S. Naval Officer Corps, 1794–1815.* Annapolis, MD: Naval Institute Press, 1991.

McKenzie, Matthew G. *Clearing the Coastline: The Nineteenth-Century Ecological and Cultural Transformation of Cape Cod.* Hanover, NH: University Press of New England, 2010.

McPherson, James M. *War on the Waters: The Union and Confederate Navies, 1861–65.* Chapel Hill: University of North Carolina Press, 2012.

Melville, George W. *In the Lena Delta.* Boston: Houghton, Mifflin, 1885.

Melville, Herman. *Mardi and a Voyage Hither.* Evanston, IL: Northwestern University Press, 1970.

———. *Moby-Dick; or, The Whale.* In *Redburn, White-Jacket, Moby-Dick.* New York: Library of America, 1983.

Merk, Frederick. *The Oregon Question: Essays in Anglo-American Diplomacy and Politics.* Cambridge, MA: Belknap Press of Harvard University Press, 1967.

Miller, Edward S. *War Plan Orange: The U.S. Strategy to Defeat Japan, 1897–1945.* Annapolis, MD: Naval Institute Press, 1991.

Mills, Eric. *Biological Oceanography: An Early History, 1870–1960.* Ithaca, NY: Cornell University Press, 1989.

Mooney, Booth. *Mr. Speaker: Four Men Who Shaped the United States House of Representatives*. Chicago: Follett, 1964.

Morgan, H. Wayne, ed. *Making Peace with Spain: The Diary of Whitelaw Reid, September–December, 1898*. Austin: University of Texas Press, 1965.

Morison, Samuel Eliot. *Aleutians, Gilberts, and Marshalls, June 1942–April 1944*. Vol. 7 of *History of United States Naval Operations in World War II*. Edison, NJ: Castle Books, 2001.

———. *"Old Bruin:" Commodore Matthew C. Perry, 1794–1858*. Boston: Little, Brown, 1967.

Morrison, Dane. *True Yankees: The South Seas and the Discovery of American Identity, 1784–1844*. Baltimore, MD: Johns Hopkins University Press, 2014.

Morrissey, Katherine G. *Mental Territories: Mapping the Inland Empire*. Ithaca, NY: Cornell University Press, 1997.

Nash, Gary B. *The Urban Crucible: Northern Seaports and the Origins of the American Revolution*. Cambridge, MA: Harvard University Press, 1986.

Nash, Roderick. *Wilderness and the American Mind*. New Haven, CT: Yale University Press, 1967.

Nelson, Megan Kate. *Ruin Nation: Destruction and the American Civil War*. Athens: University of Georgia Press, 2012.

Obeyesekere, Gananath. *Cannibal Talk: The Man-Eating Myth and Human Sacrifice in the South Seas*. Berkeley: University of California Press, 2005.

O'Connell, Robert L. *Sacred Vessels: The Cult of the Battleship and the Rise of the U.S. Navy*. Boulder, CO: Westview Press, 1991.

Oliver, James. *Wreck of the Glide, with an Account of Life and Manners at the Fiji Islands*. Boston: William D. Ticknor, 1846.

Palmer, James. *Thulia: A Tale of the Antarctic*. New York: Samuel Coleman, 1843.

Paolino, Ernest N. *The Foundations of the American Empire*. Ithaca, NY: Cornell University Press, 1973.

Pastore, Christopher L. *Between Land and Sea: The Atlantic Coast and the Transformation of New England*. Cambridge, MA: Harvard University Press, 2014.

Pearson, Chris, Peter Coates, and Tim Cole, eds. *Militarized Landscapes: From Gettysburg to Salisbury Plain*. New York: Bloomsbury, 2010.

Perl-Rosenthal, Nathan. *Citizen Sailors: Becoming American in the Age of Revolution*. Cambridge, MA: Harvard University Press, 2015.

Petto, Christine Marie. *Mapping and Charting in Early Modern England and France: Power, Patronage, and Production*. London: Lexington Books, 2015.

Philbrick, Nathaniel. *In the Heart of the Sea: The Tragedy of the Whaleship* Essex. New York: Penguin Books, 2001.

———. *Sea of Glory: America's Voyage of Discovery; The U.S. Exploring Expedition, 1838–1842*. New York: Penguin Books, 2003.

Philbrick, Thomas. *James Fenimore Cooper and the Development of American Sea Fiction*. Cambridge, MA: Harvard University Press, 1961.

Pinsel, Marc. I. *150 Years of Service on the Seas: A Pictorial History of the U.S. Oceanographic Office from 1830 to 1980.* 2 vols. Washington, DC: Government Printing Office, 1982.

Poe, Edgar Allan. "MS. Found in a Bottle." In *Complete Stories and Poems of Edgar Allan Poe.* New York: Doubleday, 1966, 148–155.

Ponko, Vincent. *Ships, Seas, and Scientists: U.S. Naval Exploration and Discovery in the Nineteenth Century.* Annapolis, MD: Naval Institute Press, 1974.

Porter, David. *Journal of a Cruise Made to the Pacific Ocean by Captain David Porter in the United States Frigate Essex, in the Years 1812, 1813, and 1814.* Vol. 1. Philadelphia: Bradford and Inskeep, 1815.

Pratt, Mary Louise. *Imperial Eyes: Travel Writing and Transculturation.* London: Routledge, 1992.

Pursell, Carroll. *The Machine in America: A Social History of Technology.* Baltimore, MD: Johns Hopkins University Press, 1995.

Read, George H. *The Last Cruise of the Saginaw.* Boston: Houghton Mifflin, 1912.

Rediker, Marcus. *The Amistad Rebellion: An Atlantic Odyssey of Slavery and Freedom.* New York: Viking, 2012.

———. *Between the Devil and the Deep Blue Sea: Merchant Seamen, Pirates, and the Anglo-American Maritime World, 1700–1750.* New York: Cambridge University Press, 1987.

Reidy, Michael S. *Tides of History: Ocean Science and Her Majesty's Navy.* Chicago: University of Chicago Press, 2008.

Reidy, Michael S., Gary Kroll, and Erik M. Conway. *Exploration and Science: Social Impact and Interaction.* Santa Barbara: ABC-CLIO, 2007.

Reingold, Nathan, ed. *The Papers of Joseph Henry.* Vol. 3. Washington, DC: Smithsonian Institution Press, 1972.

———, ed. *Science in Nineteenth Century America: A Documentary History.* Hill and Wang: New York, 1964.

Remey, George C. *Life and Letters of Rear Admiral George Collier Remey, United States Navy, 1841–1928.* Edited by Charles Mason Remey. 9 vols. Washington, DC, 1939.

Rentfrow, James C. *Home Squadron: The U.S. Navy on the North Atlantic Station.* Annapolis, MD: Naval Institute Press, 2014.

Reynolds, Jeremiah N. *Address on the Subject of a Surveying and Exploring Expedition to the Pacific Ocean and South Seas.* New York: Harper, 1836.

———. *Exploring Expedition. Correspondence between J.N. Reynolds and the Hon. Mahlon Dickerson* (New York, 1838).

———. *Pacific and Indian Oceans: The South Sea Surveying and Exploring Expedition, Its Inception, Progress, and Objects.* New York: Harper, 1841.

———. *Voyage of the United States Frigate Potomac.* New York: Harper, 1835.

Reynolds, William. *The Private Journal of William Reynolds, United States Exploring Expedition, 1838–1842.* Edited by Nathaniel Philbrick and Thomas Philbrick. New York: Penguin Books, 2004.

Riley, James. *Authentic Narrative of the Loss of the Brig Commerce*. New York: Collins, 1818.

Ritchie, G. S. *The Admiralty Chart: British Naval Hydrography in the Nineteenth Century*. New York: American Elsevier, 1967.

Ritvo, Harriet. *The Platypus and the Mermaid and Other Figments of the Classifying Imagination*. Cambridge, MA: Harvard University Press, 1997.

Robinson, Michael F. *The Coldest Crucible: Arctic Exploration and American Culture*. Chicago: University of Chicago Press, 2006.

Roland, Alex, W. Jeffrey Bolster, and Alex Keyssar. *The Way of the Ship: America's Maritime History Reenvisioned, 1600–2000*. Hoboken, NJ: John Wiley, 2008.

Rouleau, Brian. *With Sails Whitening Every Sea: Mariners and the Making of an American Maritime Empire*. Ithaca, NY: Cornell University Press, 2014.

Rozwadowski, Helen. *Fathoming the Ocean: The Discovery and Exploration of the Deep Sea*. Cambridge, MA: Harvard University Press, 2005.

———, and David van Keuren, eds. *The Machine in Neptune's Garden: Historical Perspectives on Technology and the Marine Environment*. Sagamore Beach, MA: Science History, 2004.

Russell, Edmund. *War and Nature: Fighting Humans and Insects with Chemicals from World War I to Silent Spring*. New York: Cambridge University Press, 2001.

Sachs, Aaron. *The Humboldt Current: Nineteenth-Century Exploration and the Roots of American Environmentalism*. New York: Penguin, 2006.

Said, Edward. *Culture and Imperialism*. New York: Vintage, 1993.

Schlee, Susan. *The Edge of an Unfamiliar World: A History of Oceanography*. New York: E.P. Dutton, 1973.

Schley, Winfield Scott. *Forty-Five Years under the Flag*. New York: D. Appleton, 1904.

Schneller, Robert J., Jr. *A Quest for Glory: A Biography of Rear Admiral John A. Dahlgren*. Annapolis, MD: Naval Institute Press, 1996.

Schroeder, John H. *Shaping a Maritime Empire: The Commercial and Diplomatic Role of the American Navy, 1829–1861*. Westport, CT: Greenwood Press, 1985.

Schroeder, Seaton. *A Half Century of Naval Service*. New York: D. Appleton and Company, 1922.

Schulten, Susan. *The Geographical Imagination in America, 1880–1950*. Chicago: University of Chicago Press, 2001.

———. *Mapping the Nation: History and Cartography in Nineteenth-Century America*. Chicago: University of Chicago Press, 2012.

Shufeldt, Robert Wilson. *The Relation of the Navy to the Commerce of the United States*. Washington: J.L. Ginck, 1878.

Seager, Robert, II. *Alfred Thayer Mahan: The Man and His Letters*. Annapolis, MD: Naval Institute Press, 1977.

Seager, Robert, II, and Doris D. Maguire, eds. *Letters and Papers of Alfred Thayer Mahan*. 3 vols. Annapolis, MD: Naval Institute Press, 1975.

Semmes, Raphael. *Memoirs of Service Afloat during the War between the States*. Baltimore, MD: Kelly, Piet, 1869.

Short, John Rennie. *Cartographic Encounters: Indigenous Peoples and the Exploration of the New World*. London: Reaktion Books, 2009.

Shulman, Mark R. *Navalism and the Emergence of American Sea Power, 1882–1893*. Annapolis, MD: Naval Institute Press, 1995.

Sides, Hampton. *In the Kingdom of Ice: The Grand and Terrible Polar Voyage of the USS* Jeanette. New York: Doubleday, 2014.

Skelton, William B. *An American Profession of Arms: The Army Officer Corps, 1784–1861*. Lawrence: University Press of Kansas, 1992.

Sloan, Edward W. *Benjamin Franklin Isherwood, Naval Engineer: The Years as Engineer-in-Chief, 1861–1869*. Annapolis, MD: United States Naval Institute, 1965.

Slotkin, Richard. *Regeneration through Violence: The Mythology of the American Frontier, 1600–1860*. Middletown, CT: Wesleyan University Press, 1973.

Slotten, Hugh Richard. *Patronage, Practice, and the Culture of American Science: Alexander Dallas Bache and the U.S. Coast Survey*. New York: Cambridge University Press, 1994.

*Songs of the Yosemite*. Detroit: John F. Eby, 1901.

Smith, Bernard. *European Vision and the South Pacific, 1768–1850: A Study in the History of Art and Ideas*. Oxford: Clarendon Press, 1960.

Smith, Henry Nash. *Virgin Land: The American West as Symbol of Myth*. Cambridge, MA: Harvard University Press, 1978.

Smith, Neil. *American Empire: Roosevelt's Geographer and the Prelude to Globalization*. Berkeley: University of California Press, 2003.

———, and Anne Godlewska, eds. *Geography and Empire*. Cambridge, MA: Blackwell, 1994.

Spector, Ronald H. *Admiral of the New Empire: The Life and Career of George Dewey*. Baton Rouge: Louisiana State University Press, 1974.

———. *Professors of War: The Naval War College and the Development of the Naval Profession*. Newport, RI: Naval War College Press, 1977.

Sprout, Harold, and Margaret Sprout. *The Rise of American Naval Power, 1774–1918*. Princeton, NJ: Princeton University Press, 1939.

Stachurski, Richard. *Longitude by Wire: Finding North America*. Columbia, SC: University of South Carolina Press, 2009.

Stanton, William. *The Great United States Exploring Expedition, 1838–1842*. Berkeley: University of California Press, 1975.

Steinbeck, Elaine, and Robert Wallsten, eds. *Steinbeck: A Life in Letters*. New York: Penguin Books, 1976.

Steinbeck, John. *The Log from the Sea of Cortez*. New York: Penguin Books, 1995.

Steinberg, Philip E. *The Social Construction of the Ocean*. New York: Cambridge University Press, 2001.

Still, William N., John M. Taylor, and Norman C. Delaney. *Raiders and Blockaders: The American Civil War Afloat*. Washington, DC: Brassey's, 1998.

Sumida, John Tetsuro. *Inventing Grand Strategy and Teaching Command: The Classic Works of Alfred Thayer Mahan Reconsidered.* Baltimore, MD: Johns Hopkins University Press, 1997.

Tansill, Charles Callan. *Diplomatic Relations between the United States and Hawaii, 1885–1889.* New York: Fordham University Press, 1940.

Taylor, E. G. R. *The Haven-Finding Art: A History of Navigation from Odysseus to Captain Cook.* New York: American Elsevier, 1971.

Taylor, Joseph R. *Making Salmon: An Environmental History of the Northwest Fisheries Crisis.* Seattle: University of Washington Press, 1999.

Theberge, Albert E. *The Coast Survey, 1807–1867.* Vol. 1. Silver Spring, MD: National Oceanic and Atmospheric Administration, 1998. http://www.lib.noaa.gov /noaainfo/heritage/coastsurveyvol1/CONTENTS.html. Accessed April 13, 2012.

Thoreau, Henry David. *Cape Cod.* New York: Penguin Books, 1987.

Tocqueville, Alexis. *Democracy in America.* 3rd ed. New York: George Adlard, 1839.

Trask, David F. *The War with Spain in 1898.* New York: Macmillan, 1981.

Tuan, Yi-Fu. *Space and Place: The Perspective of Experience.* Minneapolis: University of Minnesota Press, 1977.

Tucker, Richard P., and Edmund Russell. *Natural Enemy, Natural Ally: Toward an Environmental History of Warfare.* Corvallis: University of Oregon Press, 2004.

Turk, Richard W. *The Ambiguous Relationship: Theodore Roosevelt and Alfred Thayer Mahan.* New York: Greenwood Press, 1987.

Twain, Mark. *The Autobiography of Mark Twain.* Edited by Harriet Elinor Smith. Vol. 1. Berkeley: University of California Press, 2010.

———. *Following the Equator: A Journey around the World.* New York: Doubleday & McClure, 1897.

Tyler, David B. *The Wilkes Expedition: The First United States Exploring Expedition, 1838–1842.* Philadelphia: American Philosophical Society, 1968.

Van Tilburg, Hans Konrad. *A Civil War Gunboat in Pacific Waters: Life on Board USS Saginaw.* Gainesville: University Press of Florida, 2010.

Verne, Jules. *Twenty Thousand Leagues under the Sea.* New York: Butler Brothers, 1887.

Vickers, Daniel. *Farmers and Fishermen: Two Centuries of Work in Essex County, Massachusetts, 1630–1850.* Chapel Hill: University of North Carolina Press, 1994.

Vickers, Daniel, with Vince Walsh. *Young Men and the Sea: Yankee Seafarers in the Age of Sail.* New Haven, CT: Yale University Press, 2005.

Viola, Herman J., and Carolyn Margolis, eds. *Magnificent Voyagers: The U.S. Exploring Expedition, 1838–1842.* Washington, DC: Smithsonian Institution Press, 1985.

Walden, Daniel. *The Coastal Frontier and the Oceanic Wilderness in "The Narrative of Arthur Gordon Pym."* Baltimore, MD: Edgar Allan Poe Society, 2011.

Wallis, Mary. *Life in Feejee; or, Five Years among the Cannibals*. Boston: William Heath, 1851.

Walls, Laura Dassow. *The Passage to Cosmos: Alexander von Humboldt and the Shaping of America*. Chicago: University of Chicago Press, 2009.

Ward, R. Gerard, ed. *American Activities in the Central Pacific, 1790–1870*. Vol. 2. Ridgewood, NJ: Gregg Press, 1967.

Weber, Gustavus A. *The Hydrographic Office: Its History, Activities and Organization*. Baltimore, MD: Johns Hopkins Press, 1926.

Weir, Gary. *An Ocean in Common: American Naval Officers, Scientists, and the Ocean Environment*. College Station: Texas A&M Press, 2001.

White, Richard. *The Organic Machine: The Remaking of the Columbia River*. New York: Hill and Wang, 1995.

Whitman, Walt. *Leaves of Grass*. New York: Bantam, 2004.

Wiley, Henry A. *An Admiral from Texas*. Garden City, NY: Doubleday, Doran, 1934.

Wilkes, Charles. *Autobiography of Rear Admiral Charles Wilkes, U.S. Navy, 1798–1877*. Edited by William James Morgan, David B. Tyler, Joe L. Leonhart, and Mary F. Loughlin. Washington, DC: Department of the Navy, 1978.

———. *Hydrography*. Vol. 23. Philadelphia: C. Sherman, 1861.

———. *Narrative of the United States Exploring Expedition during the Years 1838, 1839, 1840, 1841, and 1842*. 5 vols. Philadelphia: Lea and Blanchard, 1845.

Williams, Frances Leigh. *Matthew Fontaine Maury: Scientist of the Sea*. New Brunswick, NJ: Rutgers University Press, 1963.

Williams, Thomas, and James Calvert. *Fiji and the Fijians*. Edited by George Stringer Rowe. New York: D. Appleton, 1859.

Winichakul, Thongchai. *Siam Mapped: A History of the Geo-Body of a Nation*. Honolulu: University of Hawaii Press, 1994.

Worster, Donald. *Nature's Economy: A History of Ecological Ideas*. 2nd ed. New York: Cambridge University Press, 1994.

Wroth, Lawrence C. *The Way of a Ship: An Essay on the Literature of Navigation Science along with Some American Contributions to the Art of Navigation, 1519–1802*. Edited by John B. Hattendorf. Providence, RI: John Carter Brown Library, 2011.

## Articles and Book Chapters

Adler, Antony. "The Capture and Curation of the Cannibal 'Vendovi': Reality and Representation of a Pacific Frontier." *Journal of Pacific History* 49 (3): 255–82.

Ammen, Daniel. "Proceedings in the General Session of the Canal Congress." *Journal of the American Geographical Society of New York* 11 (1879): 153–85.

Bates, Charles C., George Tselepis, and Daniel Von Nieda. "Needed: Shallow Thinking." *United States Naval Institute Proceedings* 94 (November 1968): 43–51.

Belknap, George E. "Something about Deep-Sea Sounding," In *The United Service*. Vol. 1. Philadelphia: Lewis R. Hamersly, 1879, 161–83, 347–73.

Bernadou, J. B. "The 'Winslow' at Cardenas." *Century Illustrated Monthly* 57 (1898–99): 698–706.

Bolster, W. Jeffrey. "Opportunities in Marine Environmental History." *Environmental History* 11 (July 2006): 567–97.

Bradford, Royal B. "Coaling-Stations for the Navy." *Forum* 26 (February 1899): 732–47.

Brady, Lisa M. "The Wilderness of War: Nature and Strategy in the American Civil War." *Environmental History* 10 (July 2005): 421–47.

Braisted, William R. "The Philippine Naval Base Problem, 1898–1909." *Mississippi Valley Historical Review* 41 (June 1954): 21–40.

———. "The United States Navy's Dilemma in the Pacific, 1906–1909." *Pacific Historical Review* 26 (August 1957): 235–44.

Brown, Chandos Michael. "A Natural History of the Gloucester Sea Serpent: Knowledge, Power, and the Culture of Science in Antebellum America." *American Quarterly* 42 (September 1990): 402–36.

Buhl, Lance C. "Maintaining 'An American Navy,' 1865–1889." In *In Peace and War: Interpretations of American Naval History*, edited by Kenneth Hagan, 145–73. Westport, CT: Greenwood Press, 1978.

———. "Mariners and Machines: Resistance to Technological Change in the American Navy, 1865–1869." *Journal of American History* 61 (December 1974): 703–27.

Burnett, D. Graham. "Hydrographic Discipline among the Navigators: Charting an 'Empire of Commerce and Science' in the Nineteenth-Century Pacific." In *The Imperial Map: Cartography and the Master of Empire*, edited by James R. Akerman. Chicago: University of Chicago Press, 2009, 185–259.

———. "Matthew Fontaine Maury's 'Sea of Fire': Hydrography, Biogeography, and Providence in the Tropics." In *Tropical Visions in the Age of Empire*, edited by Felix Driver and Luciana Martins. Chicago: University of Chicago Press, 2005, 113–134.

Burstyn, Harold L. "Seafaring and the Emergence of American Science." In *The Atlantic World of Robert G. Albion*, edited by Benjamin W. Labaree. Middletown, CT: Wesleyan University Press, 1975.

Cecil, Lamar J. R. "Coal for the Fleet That Had to Die." *American Historical Review* 69 (July 1964): 990–1005.

Coates, Peter Tim Cole, Marianna Dudley, and Chris Pearson. "Defending Nation, Defending Nature? Militarized Landscapes and Military Environmentalism in Britain, France, and the United States." *Environmental History* 16 (July 2011): 456–91.

Cronon, William. "The Trouble with Wilderness; or, Getting Back to the Wrong Nature." *Environmental History* 1 (January 1996): 7–28.

Davis, Charles H. Review of "Narrative of the United States Exploring Expedition." *The North American Review* 61 (July 1845): 54–107.

Dening, Greg. "Deep Time, Deep Spaces: Civilizing the Sea." In *Sea Changes: Historicizing the Sea*, edited by Bernhard Klein and Gesa Mackenthun. New York: Routledge, 2004, 13–35.

Dodge, Ernest S. "Fiji Trader." *Proceedings of the Massachusetts Historical Society* 78 (1966): 3–19.

Drummond, Scott E. "The Nautical Chart—It's What You Make It." *United States Naval Institute Proceedings* 96 (April 1970): 116–19.

Dupré, John. "Are Whales Fish?" In *Folkbiology*, edited by Douglas L. Medin and Scott Atran, 461–76. Cambridge, MA: MIT Press, 1999.

Dupree, A. Hunter. "Christianity and the Scientific Community in the Age of Darwin." In *God and Nature: Historical Essays on the Encounter between Christianity and Science*, edited by David C. Lindberg and Ronald L. Numbers, 351–68. Berkeley: University of California Press, 1986.

Ehrenberg, Ralph E., John A. Wolter, and Charles A. Burroughs. "Surveying and Charting the Pacific Basin." In *Magnificent Voyagers: The U.S. Exploring Expedition, 1838–1842.*, edited by Herman J. Viola and Carolyn Margolis, 164–87. Washington, DC: Smithsonian Institution Press, 1985.

Gates, W. B. "Cooper's Crater and Two Explorers." *American Literature* 23 (May 1951): 243–46.

———. "Cooper's Sea Lions and Wilkes' Narrative." *PMLA* 65 (December 1950): 1069–75.

Glenn, Barbara. "Melville and the Sublime in *Moby-Dick*." *American Literature* 48 (May 1976): 165–82.

Glover, R. O. " 'Hydro' Charts a War." *United States Naval Institute Proceedings* 73 (January 1947): 27–37.

Goodrich, Caspar G. "The St. Louis' Cable-Cutting." *United States Naval Institute Proceedings* 26 (March 1900): 157–66.

Halsey, W. F. "The Last Naval Engagement of the War." *United States Naval Institute Proceedings* 24 (September 1898): 53–63.

Hardy, Penelope K. "Matthew Fontaine Maury: Scientist." *International Journal of Maritime History* 28 (May 2016): 402–10.

Igler, David. "On Coral Reefs, Volcanoes, Gods, and Patriotic Geology; or, James Dwight Dana Assembles the Pacific Basin." *Pacific Historical Review* 79 (February 2010): 23–49.

Jewell, Theodore F. "Deep Sea Sounding." *United States Naval Institute Proceedings* 4 (1877): 37–63.

Karp, Matthew. "Slavery and Sea Power: The Navalist Impulse in the Antebellum South." *Journal of Southern History* 77 (May 2011): 283–324.

Keller, Tait. "The Mountains Roar: The Alps during the Great War." *Environmental History* (April 2009): 254–74.

Kennedy, Paul M. "Imperial Cable Communications and Strategy, 1870–1914." *English Historical Review* 86 (October 1971): 728–52.

Langley, Harold D. "Winfield S. Schley and Santiago: A New Look at an Old Controversy." In *Crucible of Empire: The Spanish-American War and Its Aftermath*, edited by James C. Bradford, 69–101. Annapolis, MD: Naval Institute Press, 1993.

Leighly, John. Introduction to *The Physical Geography of the Sea and Its Meteorology*, by Matthew Fontaine Maury. Cambridge, MA: Harvard University Press, 1963, ix-xxx.

Lindgren, James. "That Every Mariner May Possess the History of the World": A Cabinet for the East India Marine Society of Salem." *New England Quarterly* 68 (June 1995): 179–205.

Livermore, Seward W. "American Naval-Base Policy in the Far East." *Pacific Historical Review* 13 (June 1944): 113–35.

———. "Theodore Roosevelt, the American Navy, and the Venezuelan Crisis of 1902–03." *American Historical Review* 51 (April 1946): 452–71.

Mahan, Alfred T. "Blockade in Relation to Naval Strategy." *United States Naval Institute Proceedings* 21 (December 1895): 851–66.

———. "Hawaii and Our Future Sea-Power." *The Forum* (March 1893): 1–11.

———. "The Isthmus and Sea Power." *Atlantic Monthly* 72 (October 1893): 459–72.

———. "A Twentieth-Century Outlook." *Harper's Monthly Magazine* 95 (September 1897): 521–33.

Malloy, Mary. "Sailors' Souvenirs at the East India Marine Hall: 'Gathered, with Cost and Pains, from Every Clime.'" *Log of Mystic Seaport* 37 (Fall 1985): 93–103.

Maurer, John H. "Fuel and the Battle Fleet: Coal, Oil, and American Naval Strategy, 1898–1925." *Naval War College Review* 34 (November/December 1981): 60–74.

Maury, Matthew Fontaine. "Blank Charts on Board Public Cruisers." *Southern Literary Messenger* 9 (August 1843): 458–61.

———. "On the Navigation of Cape Horn." *American Journal of Science and Arts* 26 (July 1834): 54–63.

McCarty-Little, William. "The Strategic Naval War Game or Chart Maneuver." *United States Naval Institute Proceedings* 38 (December 1912): 1213–33.

McNeill, J. R. *Mosquito Empires: Ecology and War in the Greater Caribbean, 1620–1914*. New York: Cambridge University Press, 2010.

Melville, George W. "The Important Elements in Naval Conflicts." *Annals of the American Academy of Political and Social Science* 26 (July 1905): 123–36.

Nash, Linda. "The Agency of Nature or the Nature of Agency." *Environmental History* 10 (January 2005): 67–69.

Nuttall, Paul, Paul D'Arcy, and Colin Philp. "Waqa Tabu—Sacred Ships: The Fijian Drua." *International Journal of Maritime History* 26 (Fall 2014): 427–50.

Philbrick, Thomas. "Romanticism and the Literature of the Sea." In *The Eighteenth Century and the Classic Age of Sail*. Vol 2 of *Maritime History*, edited by John B. Hattendorf. Malabar, FL: Krieger, 1997.

Pinsel, Marc I. "The Wind and Current Chart Series Produced by Matthew Fontaine Maury." *Navigation* 28 (Summer 1981): 123–37.

Rager, Scott William. "Uncle Joe Cannon: The Brakeman of the House of Representatives, 1903–1911." In *Masters of the House: Congressional Leadership over Two Centuries*, edited by Roger H. Davidson, Susan Webb Hammond, and Raymond W. Smock, 63–89. Boulder, CO: Westview Press, 1998.

Reingold, Nathan. "Two Views of Maury . . . and a Third." *Isis* 55 (September 1964): 370–77.

Sahlins, Marshall. "Artificially Maintained Controversies: Global Warming and Fijian Cannibalism." *Anthropology Today* 19 (June 2003): 3–5.

Sampson, William T. "The Atlantic Fleet in the Spanish War." *Century Magazine* 57 (April 1899): 886–914.

Seager, Robert, II. "Ten Years before Mahan: The Unofficial Case for the New Navy, 1880–1890." *Mississippi Valley Historical Review* 40 (December 1953): 491–512.

Smith, Geoffrey Sutton. "The Navy Before Darwinism: Science, Exploration, and Diplomacy in Antebellum America." American Quarterly 28 (Spring 1976): 41–55.

Squier, George Owen. "The Influence of Submarine Telegraph Cables upon Military and Naval Supremacy." *United States Naval Institute Proceedings* 26 (December 1900): 599–622.

Steinberg, Theodore. "Down to Earth: Nature, Agency, and Power in History." *American Historical Review* 107 (June 2002): 798–820.

Stockton, Charles H. "The Commercial Geography of the American Inter-Oceanic Canal." *Journal of the American Geographical Society of New York* 20 (1888): 75–93.

———. "Submarine Telegraph Cables in Time of War." *United States Naval Institute Proceedings* 24 (September 1898): 451–56.

Sumida, John Tetsuro, and David Alan Rosenberg. "Machines, Men, Manufacturing, Management and Money: The Study of Navies as Complex Organizations and the Transformation of Twentieth Century Naval History." In *Doing Naval History: Essays toward Improvement*, edited by John B. Hattendorf, 25–40. Newport, RI: Naval War College Press, 1995.

Taylor, Alan. " 'Wasty Ways': Stories of American Settlement." *Environmental History* 3 (July 1998): 291–310.

Taylor, Albert Pierce. "The American Navy in Hawai'i." *Proceedings of the United States Naval Institute* (August 1927): 907–33.

Winslow, Cameron McR. "Cable-Cutting at Cienfuegos." *Century Illustrated Magazine* 57 (March 1899): 708–17.

## Dissertations

Alaniz, Rodolfo John. "Dredging Evolutionary Theory: The Emergence of the Deep Sea as a Transatlantic Site for Evolution, 1853–1876." PhD diss., University of California, San Diego, 2014.

Costello, Daniel Joseph. "Planning for War: A History of the General Board of the Navy, 1900–1914." PhD diss., Tufts University, 1968.

McKenzie, Matthew Gaston. "Vocational Science and the Politics of Independence: The Boston Marine Society, 1754–1812." PhD diss., University of New Hampshire, 2003.

Towle, Edward Leon. "Science, Commerce and the Navy on the Seafaring Frontier." PhD diss., University of Rochester, 1966.

Wijkmark, Johan. " 'One of the Most Intensely Exciting Secrets': The Antarctic in American Literature, 1820–1849." PhD diss., Karlstadt University, 2009.

# Index

Charts, 29; and authority of, 8–9, 82; commercial importance of, 3, 14–15, 21, 25, 31, 35, 39, 42, 47, 49–50, 53–54, 60, 62–63, 70–71, 75, 138, 178–80; confidential, 31, 175; and construction of, 148; and control of environment, 2, 5–6, 8, 11, 38, 42, 49–50, 208; general significance of, 2, 12; imaginative power of, 5, 8, 10, 28, 37–38, 49, 51, 62, 69, 72, 82–83, 86, 91, 100, 130, 191; and international cooperation, 8, 93; limitations of and flaws in, 3–4, 8, 10–11, 22, 64–65, 67, 95–96, 103–4, 141, 147–48, 152, 155–56, 160, 163, 169–72, 200; and method of printing, 96, 106; as narratives, 25, 30, 37, 42, 58; Navigator's Chart, 169–72; and number relative to British, 30–31, 112–13, 173; and power to claim, 58; role in courts of inquiry, 160–63; role in strategic debates, 166–69, 188; sourced from maritime community, 15, 31; strategic importance of, 3, 7, 31, 138–39, 152, 163, 173; war charts, 188–89. *See also* Cartography

Civil War, 35, 101, 105, 113, 183; and charting waters of naval operations in, 109; and *guerre de course*, 105, 107, 109–10; as watershed, 4, 6, 21, 45, 94, 97, 100, 108–9, 121

Coaling stations, 108, 138, 166–67, 172, 199–200

Cook, James, 19, 23–24, 27, 30, 43, 55–57, 59, 131

Cooper, James Fenimore, 18–19, 26, 39, 41, 69, 75

Coral reefs, 22, 40, 52, 60; meanings ascribed to, 24–25; preservation of, 1; relative to culture, 37, 42, 52, 61–62, 120, 152, 203, 206, 208

Corps of Engineers, U.S. Army, 31, 34, 109

Courts of inquiry, 125, 140–41, 145–46, 160–63, 169–72

Craig, Joseph C., commanding *Albany*, 140, 160–63; as hydrographer, 148, 175

Cuba, 135, 141, 148–49, 188; and Cienfuegos, 147, 152–53, 177; and Guantánamo Bay, 172, 177, 182, 186, 196–97; and Havana, 147, 154, 177–78, 198; protests American surveys, 199; and Santiago Bay, 147, 177, 186, 197; strategic position of, 144, 175

Dana, James Dwight, 47, 55–56, 99

Dana, Richard Henry, 16, 18

Davis, Charles Henry, 81, 133

Depot of Charts and Instruments, establishment of, 15, 28, 31–32, 34, 112; under Gillis, 36, 79; under Maury, 79; under Wilkes, 35

Dewey, George, and charting and surveying, 174; and General Board of the Navy, 174, 192, 198; on naval bases, 174; and 1903 naval exercises, 193–96; and Spanish-American War, 159, 162

Diplomacy, 12, 27, 49, 62, 81, 101, 108, 122, 138, 172, 176, 199–200

Downes, John, 21, 45

Emmons, George Foster, 48, 51, 53, 60–64, 66

Empire: and commerce, 6, 15, 17–21, 25, 28–29, 39, 42, 48–49, 52, 57, 59–60, 65, 69, 72–73, 76, 105–6, 108, 121, 200; and legacy of, 1; and naval strategy, 7, 28–29, 108, 140, 152; as precursor to tourism, 115, 127; role of charts in, 8, 39; role of science in, 2; and resource extraction, 19, 23, 91, 206; and the sea, 2, 40; territorial, 20–21, 56, 108, 111, 141, 164, 181, 183

Environment: agency of, 4, 9–11, 37, 43, 65, 76, 118–19, 148, 164, 176–77, 200; and climate change, 206–7; as data kept in ships' logs, 80; deep sea, 22, 37, 40, 75, 102–4, 120, 126–35, 187; and environmental determinism, 10; and environmentalism, 13, 206; littoral, 37, 42, 50, 52; as machine, 98; and marine organisms, 9, 14, 39, 90–91, 134; and meteorology, 14; militarization of, 108, 132, 143–45, 148, 168–69, 194; as natural enemy, 11–12, 141, 150–51, 163, 197; and naval operations, 2, 156, 200; and ocean winds and currents, 14, 29, 40, 77, 88–89, 101, 120; polar, 44–45, 56–58, 111; sea floor, 103, 106, 153; significance of, 10–13; and storms, 5, 15, 17–18, 37, 65, 80, 120, 127, 144, 159, 176, 187, 198; terrestrial, 3, 7; and tides, 5, 30, 162. *See also* Coral reefs; Sea

Fiji Islands: and cannibalism, 23–24, 37–38, 61, 66–68, 71; and Fijians, 66–67, 114, 148; and Malolo attack, 43, 67–68, 101, 116; and uncharted reefs, 22–23; and U.S. Exploring Expedition, 38, 42–43, 59–69
Franklin, Benjamin, 32, 97

General Board of the Navy, 165, 175, 193–94; role of, 174, 196; and naval base debate, 174, 182, 187, 196 98
Germany, threats of war against, 167, 181, 193, 201
Groundings, 1, 68, 140–41, 145, 151, 154, 161–63, 181
Gulf of Mexico, 98, 133–34
Gulf Stream, 23, 83, 97, 104

Hassler, Ferdinand, 34–35, 50, 97
Hawaii, American interest in, 137–39; and Mahan, 144; and role in American

empire, 56, 111; and Trans-Pacific Cable, 135. *See also* Pearl Harbor
Henry, Joseph, 79, 99, 122, 124–25, 133
Humboldt, Alexander von, 43, 45, 55, 83, 100
Hydrography, 3; defined as, 2; language of, 168–69. *See also* Charts

*Influence of Sea Power upon History, 1660–1783, The,* 142–43, 187

Japan, 56, 101, 107, 128, 130, 132, 198, 206
*Journal of a Cruise Made to the Pacific Ocean,* 26

Mahan, Alfred Thayer, 3, 5, 145, 153, 166, 186, 204, 209; and blockade, 146; and hydrography, 144–45; and Naval War Board, 146, 174; and Naval War College, 142, 187; and sea power, 2, 5, 7, 11, 27, 29, 142–43, 153, 162, 164–65, 168, 173, 175–76, 197, 199, 201; as writer, 143–44, 201. See also *Influence of Sea Power upon History, 1660–1783, The*
Manifest Destiny, 20, 49, 69, 84, 94, 136
Maps, 8–9, 12, 25–26, 30, 39, 65, 87, 121
Marine Environment. *See* Environment
Mariners, dangers facing, 15, 22–23, 61, 76 77; and empire, 18 19; and encountering the sea, 15–17; folkloric beliefs of, 12, 14–15, 17, 82, 84, 105, 119; as laborers, 16–17, 25; and literacy, 25, 40; and literature, 19, 22–26, 44; and navigational practices of, 16, 43–44; as observers, 14, 77, 85–86, 94–95, 100; scientific understandings of, 6, 12, 14, 17, 77, 95; value of charts to, 4, 14–15; views on outsiders, 18, 94; views on wilderness, 37

CPSIA information can be obtained
at www.ICGtesting.com
Printed in the USA
LVHW092049120821
695154LV00013B/1792

9 781469 659220